What Price for Privatization?

What Price for Privatization?

Cultural Encounter with Development Policy on the Zambian Copperbelt

Elizabeth C. Parsons

LEXINGTON BOOKS
A division of
ROWMAN & LITTLEFIELD PUBLISHERS, INC.
Lanham • Boulder • New York • Toronto • Plymouth, UK

Published by Lexington Books
A division of Rowman & Littlefield Publishers, Inc.
A wholly owned subsidiary of The Rowman & Littlefield Publishing Group, Inc.
4501 Forbes Boulevard, Suite 200, Lanham, Maryland 20706
http://www.lexingtonbooks.com

Estover Road, Plymouth PL6 7PY, United Kingdom

British Library Cataloguing in Publication Information Available

Library of Congress Cataloging-in-Publication Data

Parsons, Elizabeth.
What price for privatization? : cultural encounter with development policy on the Zambian copperbelt / Elizabeth C. Parsons.
p. cm.
Includes bibliographical references and index.
ISBN: 978-0-7391-4062-8 (cloth : alk. paper)
1. Copper industry and trade—Zambia—History. 2. Privatization—Social aspects—Zambia. 3. Zambia—Politics and government. I. Title.
HD9539.C6P37 2010
338.2'743096894—dc22 2010013987

The paper used in this publication meets the minimum requirements of American National Standard for Information Sciences—Permanence of Paper for Printed Library Materials, ANSI/NISO Z39.48-1992.

Printed in the United States of America

For nine miners who died during the course of this research:
three in Mindola on July 9, 2004;
six in Mufulira on May 18, 2005;

and

in gratitude for the life of
Steve de Gruchy.

Apene panshi pali ikalila, amabwe ayatondolo, ayabukata ifwe twaya yatendeka, twayabalalmuna.

The underground world lived peacefully with its divine rocks until we provoked and frustrated them.

—Charles, Kitwe area miner

♀

When I went to Zambia in 1984 it was on a World Bank Funded Project to rehabilitate the mines. The World Bank was lending $300 million to upgrade run-down operations and insisted on an audit and action plan to ensure the capital expenditure achieved its purpose. There was no indoctrination of the team as to Zambian culture or government objectives. For the first year 75% of our contact was with expats and many of those were incompetent. It took at least three years before I got to know Zambians and absorb even a little bit of their society. At a start we viewed our job as technological and not cultural. There was no attempt to look at Zambian culture to see how best to achieve our objectives. The programs we instituted were the same we would have done in mines anywhere. Retrospectively we did not achieve much, and we could have done far better if we had taken a longer-term view and focused on training. In fact it was in the multi-year projects where we were most successful. The World Bank staff and consultants spent even less time there and were equally poorly trained to figure out how to get the job done in Zambia.

—Independent technical consultant assigned to ZCCM privatization project

Contents

Figures

Tables

Abbreviations

AAC	Anglo American Corporation
BCE	Before the Common Era
BGRIMM	Beijing General Research Institute of Mining and Metallurgy
BSAC	British South Africa Company
CBU	Copperbelt University
CEC	Copperbelt Energy Corporation
CEO	Chief Executive Officer
DRC	Democratic Republic of Congo
FBO	Faith-based Organization
FDI	Foreign Direct Investment
GDP	Gross Domestic Product
GPS	Global Positioning System
HDI	Human Development Index
HIPC	Highly Indebted Poor Countries
HIV/AIDS	Human Immunodeficiency Virus/Acquired Immunodeficiency Syndrome
IMF	International Monetary Fund
JORC	Joint Ore Resources Committee
KCM	Konkola Copper Mines
KDMP	Konkola Deep Mining Project
KK	Kenneth Kaunda
LSE	London Stock Exchange

LuSE	Lusaka Stock Exchange
MCM	Mopani Copper Mines
MDG	Millennium Development Goal
MEF	Mindolo Ecumenical Foundation
MMD	Movement for Multi-party Democracy
MSD	Mines Safety Department
MUZ	Mine Workers Union of Zambia
NCCM	Nchanga Consolidated Copper Mines
NERP	New Economic Recovery Program
NFC	Nonferrous Metal Mining Corporation
NGO	Non-governmental Organization
PIC	Person-in-Charge
RAMCOZ	Roan Antelope Mining Company of Zambia
RCM	Roan Consolidated Mines
RST	Rhodesian Selection Trust/Roan Selection Trust
SAMREC	South African Code for Reporting of Mineral Resources and Mineral Reserves
SAP	Structural Adjustment Program
SDA	Seventh Day Adventist
TAZARA	Tanzania-Zambia Railway Authority
UCZ	United Church of Zambia
UDI	Unilateral Declaration of Independence
UK	United Kingdom
UMCB	United Missions in the Copperbelt

UNDP	United Nations Development Program
UNIP	United National Independence Party
USAID	United States Agency for International Development
USD	United States Dollar
USSR	Union of Soviet Socialist Republics
ZAMEFA	Zambia Metal Fabricators
ZAMTEL	Zambia Telecommunications Company
ZANACO	Zambia National Commercial Bank
ZCCM	Zambia Consolidated Copper Mines
ZCCM-IH	Zambia Consolidated Copper Mines—Investment Holdings
ZESCO	Zambia Electric Supply Company
ZMK	Zambian Kwacha
ZNBC-TV	Zambia National Broadcasting Company
ZPA	Zambia Privatisation Agency

♀

Foreword

An African proverb tells us "the stranger only sees what he knows." When we venture outside the familiar, we interpret new surroundings and behaviors through a cultural lens. Our lens is crafted by our experience of God, nature and society. It narrows our field of vision, influences our assumptions and blurs the line between judging and being judgmental.

Economic development is the domain of strangers. We know what seems to have worked for us back home and it is natural for this knowledge to shape the development theories and approaches we design and implement. Of course there are consultations with "stakeholders" such as local NGOs, community leaders and government officials. These are welcome insofar as they add dimension to our understanding. However, our inquiry rarely ventures into the realm of religion and spiritual beliefs. There seems to be concern that delving into this domain can precipitate a backward slide into a thicket of worldviews steeped in superstition and pre-Western thinking.

This is regrettable. In many societies the empirical is inextricably linked with the spiritual mind. It would seem to follow that understanding people's core beliefs yields valuable insights. For years I was perplexed over the highly risky sexual behavior prevalent in African societies where HIV-AIDS was rampant. People knew about AIDS. They could recite the causes and how it spread. And yet, they routinely took enormous risks. One day as I was commenting on this anomaly my African friends explained: "Ah, you see, it is not that simple. Yes, people are aware that condoms and monogamy reduce risk. However, the physical realm is subject to the spiritual realm. If forces in the spiritual realm decide you should be afflicted, you will be afflicted. Condoms and the rest will have no effect." In sum, the spiritual trumps the empirical.

Elizabeth Parsons taught me that crossing the economic divide without understanding the cultural divide is like navigating with a flawed GPS. We need to be open to new ways of looking at issues. Through innovative field research that took her underground with Zambian miners and above ground with their families and supervisors, she found worldviews profoundly different from ours but in harmony with an eco-spiritual perspective. Their concepts of abundance and scarcity, God and nature have little congruence with market based economics. Unlike many in the West who see the earth as a vast sandbox to be shaped and acted upon by human endeavor, these end users of development are more likely to cite the African proverb: "the ax forgets, the tree remembers." They ascribe to nature a spiritual dimension that strangers would do well to appreciate.

I recently attended a talk by President Ellen Johnson Sirleaf of Liberia, the first elected female president in Africa. Holder of a Harvard MA, a former senior executive at Citibank and the United Nations Development Program, Johnson Sirleaf is the rare policymaker who has sat all around the table in development dialogues. Asked if she had a message for donors, Johnson Sirleaf remarked that it is important for donors to listen and respect the voice of the people.

She would welcome Elizabeth Parsons's work. I know I do.

Ambassador Martin Brennan (retired)
International House
University of California at Berkeley
United States Ambassador to Zambia
2002 to 2004

♀

Preface

This book represents a very personal effort to understand issues that began to puzzle me shortly after I left the United States to work as a teacher/trainer in Southern Africa. Even as I eventually conducted research for a doctoral dissertation, however, I could not state precisely where I thought the investigation should fit within the literature. I only knew that, despite considerable efforts to prepare for international and cross-cultural work, my education did not equip me for many of the realities that I ultimately experienced. What was more, I found that almost every North American or European I encountered while on assignment understood this dilemma at least to some degree. We also often shared a sense of disconnection between what we found to be meaningful and important in our African settings and what our contacts outside the continent seemed to think was meaningful and important for our settings. Despite difficulties in naming what was going on, I knew that here was a problem that really mattered. It mattered to me as a person of faith because the intellectual disjuncture had ethical implications for my work and theological implications for how I understood myself, others, and God. It also mattered because I found that the disjuncture often had a negative impact on the lives of Africans even though I had gone to the continent out of a desire to serve and be helpful.

While I began searching for any reading material I could find that might help me understand more fully what was going on, I eventually discovered that many of the best sources of information and critique originated from outside the United States. And it has only been fairly recently that a series of books, articles, and professional consultations have begun addressing what is now often identified as "religion and development."[1] It is to this emerging field of inquiry that I see this volume making the greatest contribution.

With hindsight gained from work and study experience, I can now say with conviction that for some time, much official international development policy-making has suffered from an imbalanced intellectual foundation. There are a number of reasons for this state of affairs, as tracing the history of development-related thinking makes clear. The briefest survey of frequently used words and patterns of funding would demonstrate how development professionals have thought and rethought their mission over the past several decades.[2]

But, the present work maintains that much of this rethinking has simply tinkered around the edges of a set of presuppositions upon which the entire development enterprise rests. These presuppositions concern philosophical basics inherited from the Western Enlightenment tradition. Others have eloquently enumerated these qualities and a full retracing of steps is not the immediate concern. What is of note here is the possibility that great thinkers of earlier eras never intended their work to spawn some of the results it did. In particular, they may never have anticipated that the process of scientific-rationalist thinking emerging from the Enlightenment would contribute to the sort of disjointed, inharmonious world within which much of creation now finds itself. There are, of course, rigorous and valuable defenses of the creative engagement that can exist between religion and science, or the world of intuitive knowledge and that of empiricism and rationalism. But, equally vigorous arguments declare that a separation between the scientific thinking that produced modernity and the religious thinking that tied humanity back to older ways of understanding actually underlies problems in contemporary development thinking.

Despite the excellent ferment now underway with regard to religion and development, there is much work yet to be done. There is also considerable urgency, in the context of climate change and resource wars, to get on with it.

One of the most important strategies right now may be among the most obvious. This involves simply gaining access to many different ways of thinking about the situation. For how can we change problematic ways of thinking without learning from how others might think and act? Leading voices in the effort to overhaul some of modern development policy making's most difficult attributes argue that case studies can provide helpful vantage points for such learning.[3] The extended case that constitutes this volume attempts to provide one such example. It is not meant to be an exhaustive historical or anthropological study of the Copperbelt mining situation nor even of the re-privatization process. Rather, it is intended to illustrate in concrete fashion many of the theoretical points being made by others who are studying these matters. It is also, though more speculatively, intended to push the theological dimensions of the discussion for it seems to me that a huge amount of effort, particularly on the part of American Christians (of whom I count myself one), needs to go toward examining the meaning and import of our international engagement.[4]

So this book addresses principally my fellow citizens who are theologically and philosophically inclined because, as the opening chapter makes clear, American-led thinking has historically shaped and dominated the modern age of development. Since the case closely focuses upon Zambia, however, I expect that some Africanists might find this material of interest and there is generally some anthropological, historical, and sociological importance to the information as well. Some members of the academy in North America and Europe may be unhappy with certain aspects of how the material is presented. In anticipation of this, I offer two observations. First, the nature of academic inquiry overall seems to be moving toward interdisciplinary studies which are inherently messy; at the same time there is recognition on the part of many practitioners that insightful critical thinking is desperately needed yet painfully lacking in many programs of study that prepare students for policy making positions.[5] Second, in my experience, people of European descent often feel a need to qualify our observations, shading what we say to make sure we do not appear hegemonic in analyses of ourselves and others. Yet, frequently my African colleagues do not seem nearly as hesitant in this regard when speaking about themselves and their experiences of us. Consequently, and especially since overwhelming empirical evidence testifies to the extent that Africans have lived with hegemonic policies from Europe and North America, it seems to me that how *they* talk about cross-cultural matters needs to be taken seriously. If some intellectual nuance appears to get lost in the process, perhaps some other ideas—such as those pertinent to the incredible gap my colleagues and I experienced while in the field—may be found as well.

Finally, a word should be said about the organization of this book. Because it is a case study, an attempt has been made throughout to give the information tangibility and concreteness. For policy makers and practitioners unfamiliar with the physical mining environment, a brief introduction offers a sensory image of the Copperbelt mining arena along with specifics of what the operation entails. The first chapter sets the case study's context, defining key terminology and offering a brief history of thinking that has shaped the modern age of development. This chapter also makes the argument for using a narrative approach to the subject matter. (In an attempt to keep the narrative as readable as possible, extensive supporting material is generally referenced in notes at the end of each chapter.) Readers eager to proceed straight to the story itself may wish to skip over chapter two which contains technical information on the study's data gathering process. For reasons made clear at the outset of the chapter, however, it has seemed important to include this material. Chapters three and four offer contrasting background information on the Copperbelt based on what this work terms "public" and "private" versions of the story. The heart of the contemporary case is then presented in chapters five, six, and seven, structured in a way meant to evoke something of how Zambians appeared to think through their situation. The bulk of the narrative, then, concerns general points of view expressed by Zambian Copperbelt residents. But chapter eight is devoted to views expressed by European expatriates living on the Copperbelt. Their insights, by no means exact opposites of Zambians, are extremely valuable for gaining a more comprehensive sense of the context. Other analysts have, upon occasion, noted that cross-cultural cases can miss important details when vantage points of outsiders present within the local context are ignored.[6] The final chapter offers some analysis and interpretive points meant to spur reflection and additional investigation.

Cambridge, Massachusetts
Winter 2010

Notes

1. The present work relies heavily on the wisdom of those who have gone before in this endeavor. Some of the earliest writers on the broad topic of philosophical and theological underpinnings of international development include Gilbert Rist whose late 1990s work has now gone through several editions. *The History of Development from Western Origins to Global Faith*, translated by Patrick Camiller (London: Zed Books, 1997). See

also Wolfgang Sachs, *The Development Dictionary: A Guide to Knowledge as Power* (London: Zed Books, 1992) and Majid Rahnema and Victoria Bawtree, eds., *The Post-Development Reader* (London: Zed Books, 1997). Of late, see particularly, Séverine Deneulin with Masooda Bano, *Religion in Development: Rewriting the Secular Script* (London: Zed Books, 2009); Duncan Green, *From Poverty to Power: How Active Citizens and Effective States Can Change the World* (Oxford: Oxfam, 2008); Jeffrey Haynes, *Religion and Development: Conflict or Cooperation?* (New York: Palgrave Macmillan, 2007); Katherine Marshall and Marisa Bronwyn Van Saanen, *Development and Faith: Where Mind, Heart, and Soul Work Together* (Washington, DC: The World Bank, 2007). In late 2009, an initiative from within the American Academy of Religion first began to take off in this regard with a proposal for a consultation on Religious Dimensions in International Development.

2. Kurt Allen Ver Beek has done a version of this regarding references to spirituality within development studies publications, finding that it has been essentially avoided. His article on the topic is one of the first to inspire interest in the growing field of religion and development studies. Kurt Alan Ver Beek, "Spirituality: A Development Taboo," *Development in Practice* 10, no. 1 (February 2000): 31-43.

3. Marshall and Van Saanen, *Development and Faith*; Green, *From Poverty to Power*. See also the important cross-cultural case studies of Erica Bornstein (who places her own work in a context of the struggle to understand meaning and expectations) *The Spirit of Development: Protestant NGOs, Morality and Economics in Zimbabwe* (New York: Routledge, 2003), 170; Dorothea Hilhorst, *The Real World of NGOs* (London: Zed Books, 2003); and Laurie A. Occhipinti, *Acting on Faith: Religious Development Organizations in Northwestern Argentina* (Lanham, MD: Lexington Books, 2005).

4. See a recent sociological study conducted by Robert Wuthnow and his own concerns in this regard. *Boundless Faith: The Global Outreach of American Churches* (Berkeley: University of California Press, 2009).

5. See, for example, the entire first chapter of Herman E. Daly and John B. Cobb Jr., *For the Common Good: Redirecting the Economy Toward Community, the Environment and a Sustainable Future*, 2nd ed. (Boston: Beacon Press, 1994).

6. James Ferguson, *Expectations of Modernity: Myths and Meanings of Urban Life on the Zambian Copperbelt* (Berkeley: University of California Press, 1999), 32.

Acknowledgments

Warmest thanks go first to Anthony O. Balcomb for his wise guidance and sympathetic encouragement of the original study. Steve de Gruchy likewise provided enormously helpful and candid critiques all along the way; I was looking forward to sharing this volume with him. James Cochrane and Rosalind Hackett made valuable comments on the original study as well. Mary Becchi and Paula Kline thoroughly read early drafts of the present manuscript and provided important correctives. Michael Sisskin and Victoria Koulakjian of Lexington Books, Rowman and Littlefield Publishing Group have cordially given guidance throughout the publication process. Charley Thomas provided friendship, institutional and moral support before, during, and after the production of this work. Development Cooperation Ireland gave a grant that made the field work possible and for that I am most grateful. I thank Giovanna and Martin Brennan for friendship and hospitality that unexpectedly opened the door to this case study and mostly for just being who they are. Tim Henderson personally intervened to see that the field work happened as hoped at a crucial moment when it almost did not and has remained a thoughtful instructor and faithful friend ever since. Kapya John Kaoma has been my teacher and example from the time that I went to live in Africa until now. His advice and technical assistance have been invaluable. John Kafwanka, Kennedy Kazeza, and Derrick Muwina have provided friendship and years' worth of enlightening conversations. Rogers Banda deserves praise for his singular achievements in making possible my interactions with Zambia's first two presidents. Stewart Lane has my enduring gratitude for first reorienting me after radical cultural encounter upended my worldview. The numerous Copperbelt residents who gave so generously of their time and insights remain anonymous here but they know who they are and I am forever in their debt. Finally, thanks to my two best friends: My sister, Sarah Louise Sears, produced the Copperbelt map, read through manuscript drafts, and rescued me innumerable times with her technical and artistic expertise. My husband, Lin Parsons, has cheerfully lived with and supported this project for years now, believing in me beyond what I deserve.

Any errors of fact, interpretation, or omission are my sole responsibility.

♀

Introduction

The Miner's World of Work

Probably just behind environmentalists the last people on earth to romanticize the mining profession would be miners themselves. Ecologists are rightly raising alarms about how the global economy uses mining to devastate lands, animals, and peoples while creating stupendous paper wealth for a relative few. Yet, individual men and women who have been born into mining families or for whom geology, engineering, and metallurgy are fascinating endeavors often appear completely pragmatic about what they do. If they reflect on it at all, miners frequently use terse characterizations such as the work's being "tough" or "uncivilized." Usually it is the people around them who, knowing what the job entails, describe it in awed superlatives.

The experiential aspect of mining can teach a lot that economic analyses of the industry leave out. For beyond the numbers and commodities talk, industrial mining as an activity symbolizes much about modernity itself. It embodies the physical intrusion of people and machines on the earth along with the earth's occasional response through rock slides, cave ins, and fires. Here are creation, humanity, and technology, together in a relationship far more dramatic and dangerous than industrial agriculture, the closest possible parallel.

The Copperbelt is not the prettiest part of Zambia; not much above ground would suggest it as a likely tourist destination.[1] There are rivers, but they seem like streams compared to the Zambezi with its Victoria Falls down south. There are some rolling hills, but no grand escarpments, vast forests, or herds of game as can be seen out east toward Malawi. Mostly there are scrub trees, open spaces of fields cut through by little tracks where people walk back from the roadways to their homes, huge expanses of sky, and then the cities and townships. Downtown areas need paint; the townships would greatly benefit from regular rubbish disposal. They are busy, noisy places especially compared to the countryside and the relative order and punctuality of American cities.

The mines are usually on the outskirts of towns. They are quite obvious both by their mounds of slag[2] and by their superstructures, although these represent just a fraction of what goes on underneath. The easiest mine areas to spot are those around Kitwe where the road from Ndola—the only dual carriageway in the entire country—begins to run downhill just as the land opens up. In the distance, headframes for two or three shafts and the Nkana cooling towers are by far the tallest things in sight. The Chambishi mines are also quite evident, particularly at night. The road from Chingola to Kitwe, like all Zambian roads outside of town, has no lighting. And, since there are also no businesses, strip malls, or petrol stations along the way, the only nighttime illumination drivers have—other than moon and stars when it is clear—are their own headlights and those of other vehicles passing by. It is easy to tell when Chambishi has been reached because the sky glows golden orange. The hoist plant is brilliantly illuminated, various parts of the operation quite identifiable provided one has time to look out the window on the way past. After that it is pitch dark again.

The mines operate on three shifts: day, afternoon, and night (usually called back shift). Day shift starts early. In some areas of the Copperbelt everyone knows when it is time for the men to get ready. A siren—reminiscent of old World War II movies and tornado alerts in America's midwest—sounds around 4:00 AM. Most miners seem able to adjust their sleeping schedules so that they sleep when they can and awaken when they need. By 6:00 AM, they are ready to go underground.

Regardless of shift, the procedure coming into the mine is the same: to the changing room, pick up cap lamps and hardhats, then crush in. Crushing in is when miners pass the checkpoint and give their gang or mine number to the shift boss who marks them present. Every miner has his own number and, in the old days, miners' kids knew this as well as the men themselves. Nothing moved without that number; it was good for signing off at the company stores and health clinics and is still crucial in cases of underground accidents. Management can determine who is missing by who fails to check back in after an evacuation.

Compared to the vast gold mines of South Africa, Zambia's copper mines are quite small, but it still can take awhile for men to get to their places of work each day. They go underground in cages—sometimes two and three tiers containing a few hundred men in total—that move on strict, military-like schedules. In South Africa, where the different underground levels are not necessarily connected, if a miner misses his cage as it starts down, he might as well go home. Zambian miners are a bit more fortunate; tunnels and haulage ways are often connected so a miner can walk from a different level to his station if need be.

The apparatus for the cages is what people on the surface can see. They operate on a system that winds and unwinds rope around a huge drum. Depending on the design, sometimes the giant wheel that helps coil this rope is exposed at the top of the headframe. When the wheel is spinning around, miners or supplies are traveling up and down. In some of the Copperbelt mines, these cages drop straight down a half-mile or so. From there, miners may have to walk or take small trams to get to their stations for the day. In some cases, they may take yet a second, smaller cage ride until they are well over a mile down inside the earth.

Underground around the cage areas it can be noisy and bright. People are always moving about at these points. Information and safety signs are posted and the rock surface may be painted white. It does not take much of a walk, though, to get back to where the painted surfaces stop and the dark begins. In Zambia's mines, this seems typically around the same place where water comes up from below. Sometimes there can be more than a foot of water to slog through. When a pipe burst in one of the mines a few years ago, water so completely filled up the haulage way that they had to send in divers to fix it.

The farther down the mine goes the hotter it gets. There is more heat and stress on the rock, as drillers and timbermen can testify. Timbermen install the supports that keep things from collapsing while the drilling and blasting proceed. Drillers are known as those at the sharp end of the operations. They are on the front line moving the work along. Sometimes they have to stand all day in water up to their knees holding a jackhammer, attacking the rock. They have been known to shed their overalls and work naked when that seems to help with the heat. The drilling affects the air: little particles everywhere. So sometimes the men wear cloths tied around the lower parts of their faces. It helps keep out the dust and absorbs some of their sweat.

The other things that get out into the air are the fumes and haze from blasting. Blasting happens after the drillers have put holes into the rock that other miners fill with explosives. Depending on what the company's geologists have determined about the ore body's location, drillers have different fan-like patterns they cut into the rock face. Once the holes are bored and filled with emulsion, a miner sets the fuse, again according to a strict schedule. Miners arrange the blasts to coordinate with their own retreat out of the area. They have to make sure they know how the air and ventilation are flowing so that they do not send noxious fumes out into stations still being worked. When his section of the rock is blown away, each miner who oversees blasting then reports to the next person scheduled to blast. They are supposed physically to see each other so there are no mix-ups with somebody left behind as the charge goes off. Major blasting usually happens toward the ends of the shifts. This is because it takes the earth a while to settle after being so severely shaken.

Supervisors on later shifts have to see when their men can go into these new work areas. They look at the slopes and sides of the mine and also determine if the poisonous fumes have dissipated enough to make it safe to breathe. Men charged with this responsibility have no meters they take along to register vapor levels. They use their noses and their instincts.

Even if they judge the air to be okay, there is another deadly hazard to guard against. Hanging rocks can look just like small chips of stone overhead. But, if they drop they can travel with such force that a square hit on the head, regardless of hardhat, can be fatal. Miners are supposed to take long poles and poke at these hanging rocks to make them fall down before anyone goes inside the blasted area to work. Still, it is not always possible to see where or when a rock fall might happen. Occasionally even expatriate bosses wonder about acts of God when there are peculiar rock falls. One day a group of miners was prying and prying at a rock that just would not cut loose. They went out, then came back in. When they came back in, the rock fell. A miner lost his foot to that one.

Once the hazards are out of the way and the place all supported, then it is time to clear up the rubble and haul it out to the surface. In the past, all this was done by hand with shovels and wheelbarrows. Now, there is considerable underground machinery that scoops, crushes, and hoists the ore out of the ground.

In some places huge front loaders can be seen roaring out of haulage ways, like prehistoric animals emerging from their dens, dumping loads of ore into rail cars that then transport it away. Miners who operate some of these machines can do so by using joy sticks. They stand back to the side and send the loaders into the stopes or areas where the ore has collected. This is safer for the miners and more efficient for the mining company.

The hoist is what the company's geologists and production managers watch so carefully. The amount of tonnage brought up from the night before is the first thing they look for at the beginning of each day. Before computers, mine managers would gather to review these figures in early meetings sometimes called Morning Prayers. The prayers are electronic now. Figures get typed into emails and are on manager's computer screens by 7:00 AM each day. If there has been a problem—if a machine has broken down, for instance—management will immediately know. They will be able to tell because the hoist will be low. They will also know what sort of day they are going to have.

Underground, if everything goes according to schedule, miners work for eight to ten hours a day. They do not go back onto surface for breaks because they cannot. The cage schedule is so tightly controlled that once a man is underground he has to stay there until his shift is over. Some of the men bring water bottles with them. Some may bring something to eat. Most just say they are used to the work pattern. They eat before they go down and, perhaps, again when they get home. They may, however, simply want to have a beer and go to sleep.

Meanwhile the ore they have dug and hoisted is refined and smelted and taken away. Tiny bits of Zambia must be everywhere around the globe by now: in jewelry, piping, cell phones, and computers.

♀

It may be impossible to live on the Copperbelt and not become at least curious about mining. I certainly was intrigued by the smokestacks and head frames of the urban skylines and when finally presented with a chance to go underground, eagerly took it. Our small party spent several hours touring the operation; long enough to appreciate the intense heat, fumes, and stifling humidity that are the routine work environment for thousands of Zambian men; long enough for one in the group to remark how he realized that "mining is not for sissies."

On the ride up to surface in the cage—dark except for the flashes of light from cap lamps and dripping with water—the executive leading us remarked, "When our owners made their first visit out here, I arranged for them to go underground but they wouldn't come. They said to me, 'That's why we have you there. We just look at the numbers and if they're not good enough, you're out.'" He went on, "You've now been in more mines than they have."

Notes

1. As one observer has phrased it, "Zambia arouses little enthusiasm among visitors with an eye for scenic beauty." Richard Hall, *Zambia* (London: Pall Mall Press, 1965), 1.

2. Slag is waste material accumulated from the smelting process.

Chapter One

Stories We Tell Ourselves

This case study will demonstrate that an understanding of differing worldviews is crucial to appreciating what people mean by and hope for regarding the concept of development. Moreover, the case will show how radical differences in interpretation of this concept can arise depending upon the cosmologies that constitute different people's perceptions of reality. The thesis of this work is that the ways in which cosmologies are understood affect people's everyday lives to such an extent that genuinely helpful development policy making can only be done when this circumstance is understood. Worldviews and cosmologies matter not just at a theoretical level but pragmatically and experientially.

Despite the increasing and diverse global involvement of Americans in international development initiatives,[1] inside the United States terms such as "worldviews" and "cosmologies" have only rather recently been associated with the word "development."[2] So, it is important to be quite clear about what is at stake. People working in cross-cultural settings, where so many interventions associated with the concept of development take place, can experience perceptual discrepancies that are not accounted for within general development theory or policy making but that affect what happens in practice. These discrepancies pertain to the ways in which various actors in the situations understand how the world works. Simultaneously such discrepancies involve deep seated values issues. What then happens as a result of discordant understanding makes a difference morally and theologically.[3]

Before illustrating how this might be so by considering a particular Zambian situation, some expansion of foundational ideas is necessary.

What is Meant by "Worldview?"

A tidy definition of "worldview" is that it encompasses "the assumptions people hold about the world and how it works."[4] Not everyone agrees that there is such a thing as a worldview and, as shall be explained more fully below, there is good reason to be aware of the concept's limitations. But there is also a compelling history of thinking about worldviews that applies well to how development policies are implemented and interpreted.[5]

Philosopher Stephen Pepper is sometimes credited with coining a version of the term. His 1942 book, *World Hypotheses*, put forward the idea that there can be various models of how people "attempt to provide a complete view of the world."[6] What is more, "each world hypothesis creates its own field of play. Within that field of play, 'competing' world views can be interpreted but cannot compete directly."[7] In other words, holders of one worldview can recognize to only a limited extent the ways in which holders of other viewpoints organize and interpret their worlds. Pepper identified four distinct ways of viewing the world, including two (mechanism and organicism) that are especially appropriate for the present purposes. A common metaphorical understanding in existence since at least the beginning of the industrial era sees the world like a giant machine in which problems are breakdowns to be solved by imposing standardized solutions. By contrast, organicism is inclined to perceive the world more like a being with all the contradictions and peculiarities that accompany living entities.

While this study is not intended to match Pepper's theory with the lived experiences of Copperbelt residents, distinctive traces of a mechanistic perspective were evident in the development policy making to which Zambians were subjected during the mine privatization process. Simultaneously, a fluidity of existence inherent in Pepper's notion of organicism was also detectable within the Copperbelt setting. So it is important to note the intellectual seriousness of this issue. Discrepancies between points of view and misunderstandings that emerge especially in cross-cultural encounters are not simply due to lack of good communication or inadequate training. A solid line of philosophical investigation suggests that these issues reach to the core of what it means to be human.[8]

Scholars across academic disciplines have been interested in understanding how the worldview concept might apply to entire societies and not just to individuals. They have investigated how groups of people do or do not propagate their perspectives beyond themselves and over time. Much of this work has been conducted under the rubric known as the "sociology of knowledge," which has distinctive philosophical roots. Alfred Schutz along with his disciples, Thomas Luckmann and Peter Berger, have led in melding theories about how people see the world with what these perspectives might mean for actions.[9] A key observation from their work is that everyone makes sense of the world largely without consciously thinking about it yet proceedings that flow from perceptions make a difference in the everyday world. Indeed, we all invoke images, stories, experiences, and so on that, taken together, create "maps" or "frameworks" by which we determine what is real and meaningful. This process is not orderly, sequential, or logical.[10] Nor is it confined to an abstract level. Instead, making sense of the world from an intellectual standpoint inevitably leads individuals and groups to have both emotional reactions to and invoke moral judgments on the circumstances of their worlds.[11] In short, we think, feel, and act out of, or based upon, our worldviews.

Worldviews can represent the best means for explaining why people approach life differently, the worst means of stereotyping entire societies, or something in between. Some scholars, for example, worry that labeling groups of people according to their viewpoints is a way of suggesting that their perceptual traits are in-born. They are right to caution against what is known as "essentialism" and it is particularly easy to fall into essentialist thinking when attempting cross-cultural analysis. Caution about assigning essential qualities to entire societies seems to have arisen, in part, as a reaction to the sort of mechanistic perspective that Pepper named back in the 1940s. Post-modern thinking has critiqued mechanistic viewpoints, focusing instead on the relativity or subjectivity of life. This corrective can be a valuable guard against the temptation to impose sweeping, general analyses. But, a decade into the twenty-first century, there also seems to be a growing recognition that emphasis on relativity and subjectivity can, itself, be overdone. What is sometimes known as post-post-modern thinking searches, among other things, for points at which we may find common ground among and across groups. For this reason, an approach that takes the philosophical concept of worldviews into account is a useful way of interpreting why people and institutions act as they do.

When the concept of worldview is further understood to encompass cosmologies, the situation becomes particularly complicated in relation to issues of development. This is largely due to the intellectual context that has produced the concept of development. To put it bluntly, problems arise in large part because of certain thinking patterns that have shaped the ways in which development has historically been conceived and talked about.[12] So, a few important points about the evolution of this concept need to be made.

What is Meant by "Development?"

By many accounts, "development" is a contested term.[13] But, one way or another, it appears to involve core contemporary global concerns of business, government, and civil society. For example, in business, *human-related activity since the Industrial Revolution* is now identified by almost all climate experts as being a main precipitator of climate change.[14] In response, many entrepreneurial ideas and new technologies are being put forward as offering solutions to global warming.[15] So "development" in a business sense seems to be both part of the problem that is causing climate change and a main method for mitigating damage that is anticipated to occur as the earth's atmosphere heats up. In the governance realm, the United States' foreign policy agenda relies heavily on propagating development strategies around the world. *Providing financial assistance, materials, and technical expertise* to alleviate suffering from violent conflict and other disasters in foreign nation-states is not a strategy being invoked for the first time with the present Iraq and Afghanistan wars.[16] Aid provision has long been understood as a smart way of engaging diplomatically with weak nation-states whose stability is seen as integral to this country's well-being.[17] So "development" can be understood in a political sense as constituting sets of policies and procedures that flow from self-interest but that can also include an interest in a stable international order. Within the realm of civil society, the Millennium

Development Goals (MDGs) of the United Nations, an eight-point set of objectives focused on *raising the living standards of the world's poorest populations*,[18] are at present the most prominent coordinated methods by which attention is going to the debilitating nature of extreme poverty and rampant disease that characterize life for a staggering proportion of the world's population. In part because the MDGs attempt to assess development in ways other than economic statistics[19] they have become focal points for social justice action campaigns and popular education efforts within faith communities as well as non-governmental organizations. So "development" can also be conceived as a strategy for achieving a global common welfare that, of necessity, addresses moral and ethical issues.[20]

While there is undeniable good in all such efforts to address critical global concerns, these endeavors have conceptual and operational limitations. For instance, what is likely to happen if an assumption becomes wide spread that technological solutions can and will alleviate global warming? When the physical resources and commodities required for those technological innovations are finite, what sorts of problems might then result? Similarly, given that the world appears to be a less, not more, stable place now than a decade ago, what is the appropriateness of linking aid and technical assistance to foreign policy agendas that also include heavy military involvement?[21] Yet again, given the deepening disparities between rich and poor worlds over the past several decades,[22] why do even the MDGs' comparatively modest objectives appear unattainable?[23] Despite the efforts of faith-based campaigns to raise awareness about these measures, is there more that members of civil society should be doing?[24] In all, then, it appears that development concerns are pervasive, yet not straightforward in action, definition, or ramification.

One reason the concept of development is notoriously difficult to define is the interdisciplinary nature of the subject matter. Various social sciences have differing stances on conceptions of development and especially of the non-quantifiable aspects of people's lives (such as what constitutes well-being; how moral issues are evaluated; and so on) that are implicated in development actions.[25] Another reason for definitional difficulties is that meanings have shifted substantially several times since the 1940s and 1950s as experience has been gained and thinking has evolved.[26] A third reason for difficulty in classifying development as a concept is that there is considerable disagreement among scholars as to how much emphasis should be placed on one or another definition at any given time and in any given location. Some of this hesitation undoubtedly arises from post-modern concerns that proffering one overarching viewpoint negates or denies how the situation looks from a radically different standpoint. This is an especially wise caution when development interactions take place cross-culturally.[27]

Yet, to suggest that certain points of view have not, for periods of time, been dominant within the overall context and history of the subject would also be unwise. For example, in attempting to honor experiential differences it is possible to overlook what has been termed the "economic hegemony"[28] that has characterized much development thinking since the late 1940s.[29] The present work contends that various intellectual ideas, or schools of thought, have played prominent roles in how the notion of development has generally been defined and implemented by Western (especially American) interests. These ideas can be understood as forming something of a dominant development narrative; such a narrative had profound influence over the ways in which actions unfolded during the mine privatization process in Zambia.[30] The task of the next several chapters is to illustrate in largely storied form how this came to pass.[31] But several qualities inherent in the basic narrative deserve special mention first.

Development and Materialism

From an historical standpoint, "development" defined as what happens when individuals and groups build such things as schools, hospitals, and roads has been around for quite some time. This form of understanding the concept is sometimes thought of as a materialist perspective[32] meaning that development is predominantly about creating or constructing physical things that lead to improvements in human life.[33] In the popular imagination such activity is often associated with the colonial era, particularly as entrepreneurs and religious workers from Europe and North America carried out projects in Africa, Asia, and Latin America. Those doing these things, however, would have called themselves explorers, business persons, or missionaries rather than development practitioners or experts. Building and constructing activities were sometimes regarded by those who undertook them as ancillary means by which they could obtain more sought after objectives: wealth from resource extraction or faith conversions by those living where the projects were located.[34]

For those non-Westerners upon whom such efforts were largely enacted, the situation could appear quite different, however. From a non-Western perspective understanding "development" as principally constituting the construction of buildings or obtaining vehicles and various technologies could invite a host of problems even if some benefits might also accrue. Many of the most egregious problems, particularly during the colonial era, have been quite well documented by experts in a variety of intellectual disciplines.[35]

A materialist understanding of development has not been confined to late nineteenth and early twentieth century international faith-based interventions, however. In fact, one of the driving forces behind increased material production worldwide seems to be quite opposite from the religiously motivated efforts of missionaries and faith communities generally. That is the notion that modernity, represented by the manufacture of goods and services made possible by industrialization and technology, inevitably makes faith and God obsolete.[36] Indeed, a materialist perspective plays a prominent role in defining how public conversations typically differentiate "developed" from "undeveloped" or "developing" countries.[37] The unspoken story line associated with these terms might be stated thusly: *Developed countries have an array of visible amenities that render human life more comfortable and efficient. These countries also have the wherewithal to keep such systems going. Other countries have something less than this and so may be considered "undeveloped," "under-developed," or "developing" depending upon the specific mix of their political and economic circumstances. Further, the objectives for those existing in un- or under developed states are generally seen as moving "toward some approximation of Western material and economic standards."*[38]

But it should be noted that external interventions done with this specific understanding of development in mind can easily lead to wildly inappropriate actions such as providing equipment never meant to function in the locale where it is placed; starting businesses without forethought for markets and needs where the ventures are situated; or completely overlooking local resources and problem solving techniques.[39] A materialist emphasis on development can also lead to project failure when those entrusted with maintenance of such a built environment are insufficiently trained in its upkeep.[40] Worst of all, thinking about development largely in physical or material terms without sufficiently understanding the production and maintenance mindset behind it all can keep the very persons who are ostensibly development's beneficiaries in positions of ignorance as to what leads some—and not other—societies to have a plenitude of such things.[41]

It further bears mention that the conversion agenda of colonial era Western missionaries who provided material things such as schools and hospitals for non-Westerners has been heavily critiqued within the United States (as it should be). Yet, a similar critique of the conversion aspect of much Western development policy making, however, is almost entirely absent from American-based literature on the topic.[42] Materialism, then, figures prominently in narratives that dominate much development thinking but the ways in which the material is understood vary significantly, as do analyses of its powerful influence.

Political and Economic Influences

As noted above, development aid and/or technical assistance has been an instrument of American foreign policy for many decades. This strategy is not however, an American invention. The propensity to foster, even impose, societal transformation by means of diplomacy, force, bribery or a mix of all three was also integral to the European colonial project.[43] But, for the present study, we need be most concerned with how the modern "age of development,"[44] as Wolfgang Sachs terms it, came about since its origins are tied to American politics.

As the case to come will illustrate, by the time development policies were being initiated in twenty-first century Zambia, America's role in or influence upon that country was not immediately obvious to the casual observer. But those who implemented a series of interventions and those who lived with the consequences of such interventions did exist at some level within an intellectual context implicit to the United States. So it is imperative to understand the role of the United States in helping development policies and practices become central to contemporary global concerns.[45]

Toward the end of World War II, the Allied powers sought to establish a secure world order by devising international institutions that would, among other things, guard against the types of destabilized economies European countries had experienced in the pre-war years. The Bretton Woods Agreement that resulted from a meeting held in New Hampshire created the World Bank, the International Monetary Fund, and envisioned a precursor to today's World Trade Organization.[46] Ironically, though these institutions originated from American led meetings and remain headquartered in Washington, DC, today more non-Americans than Americans seem aware of their existence and the influence they exercise globally. In the present case, the power of the World Bank over the lives of Zambians has been particularly profound.

Bretton Woods represented an unprecedented reshaping of global power structures and following the Agreement, President Harry S. Truman perhaps unwittingly helped alter the world order even further. Toward the conclusion of his 1949 inaugural address he promised that technical assistance being given to Latin America at the time would be extended through a "new program" to distribute the benefits of science and industrial progress worldwide. This was the last of four points in Truman's speech and some analysts have suggested it was included more for rhetorical and public relations purposes than anything else.[47] But the Point Four Message, as it has come to be known, seems instead to have inspired a new industry. It did this in a two-fold manner. First, it posited a different way for people in North America and Europe (or the Global North) to think about themselves in relation to the rest of the world (or the Global South): for the first time, the planet could be conceived of as divided into "developed" and "under developed" countries.[48] Second, it gave license for the people and institutions of the Global North—those inhabiting the "developed" world—to implement a host of mechanistic strategies that were tacitly understood as necessary to bringing everyone into conformity with the "developed" model.[49] The criteria for distinguishing developed from undeveloped nations being figured in monetary terms meant that the United States—its wartime leadership still a fresh memory and its post-war Gross Domestic Product (GDP) steadily rising—became the standard to which every other nation was to aspire.[50] This, then, was the context within which "development" as a modern concept came into being. Development—an economically-oriented vision implemented within the political structures of nation-states—was the means by which the traumatized post World War II world could be transformed into a place of peace and prosperity.

Over time, however, the comparative narrowness of this definition became apparent. One difficulty was that it fostered a certain way of thinking about how problems should be defined and solutions posed. Certain economic theories in particular, dominated the means by which societal dilemmas were addressed. But "economics," a social science, has a long tradition of concerning far more than monetary things. In fact, the Greek root of the term refers to comprehensive oversight or stewardship of one's household responsibilities, not simply financial management and accumulation of cash. This richer perspective, however, was almost entirely obscured within the framework of a narrative given over to statistical measurements and growth.[51] The notion of understanding the world in nation-state terms created additional difficulties since the nation-state as a political structure is a fairly recent innovation within the sweep of history. Government-to-government aid agreements that seemed to make sense according to

Truman's policies, then, turned out to have little meaning for how ordinary people in "developing" countries lived their lives.[52] Such theoretical limitations eventually became obvious to prescient analysts. But it took several decades for the lived ramifications of the policies and programs that flowed from Point Four to become obvious.

Some sixty years into the age of development, those with experience as providers and recipients of Truman's new program have identified a range of dangers and systemic problems created when "development" is defined according to measurable, financially-oriented and politically-enacted interventions. Many of these analysts have come from settings outside the United States because it is only from such an external perspective that the limitations have been so obvious. Some of their initial critiques have argued that, rather than bringing stability to the world, the Point Four Message helped precipitate the international debt crisis of the 1970s and 1980s; post-colonial instability within nation-states of the Global South; rising worldwide economic inequity; and widespread ecological degradation.[53] As these structural problems have been identified, more holistic visions of the concept have emerged, stressing thinking in terms of processes that can lead toward human flourishing.[54] It is probably not coincidental that such expanded understandings have coincided with the professionalization of service provision and development interventions through the proliferation of civil society organizations, specifically non-governmental and faith-based organizations (NGOs and FBOs). Such organizations are generally better positioned than are government and business entities to work across intellectual boundary lines and to speak credibly about aspects of existence that statistical measures cannot capture.[55] Terminology that has attempted to capture this wider vision has begun to speak of "human-centered" or "people-centered" development, "sustainable development," and "development as transformation."[56] In particular, as analysts of religion and development Jeffrey Haynes and Séverine Deneulin both observe, the notion of "human development" now tends to predominate, stressing non-quantifiable facets of people's lives.[57] But, as development expert Amartya Sen has argued, barriers that inhibit people's abilities to act independently and resourcefully must also be broken down.[58]

As the second decade of the twenty-first century commences, there seems a growing awareness that humanity can no longer be the sole focus of development initiatives. Realizations that climate change and ecological degradation transcend the abilities of political and economic discourse to mitigate them are bringing fresh attention to the limitations of all previous definitions. They are also acknowledging the possibility that religious and spiritual points of view, aspects of existence associated with cosmologies, have something profound to contribute.

Culture and Religion in Relation to Development

Speaking or writing about culture and religion as if the two concepts are separate entities is a peculiarly Western way of thinking that can lead to unfortunate encounters with non-Westerners, especially when development issues are involved.[59] While the term "religion" generally makes sense in European and North American settings, it does so because of a prevailing set of cultural and linguistic norms. These norms are traceable to seventeenth century or earlier European attempts at identifying and labeling new practices that Christian missionaries encountered when technological advances first made global travel feasible.[60] Throughout most of the world, however, "it is implausible to believe that the religious factor can be isolated from life's general context."[61] In other words, genuine human development or flourishing is an all encompassing matter that must be understood as involving the spiritual as well as the material,[62] a point of view increasingly being recognized in policy making circles. The United Nations Development Program's (UNDP) 2004 *Human Development Report* on cultural diversity, for example, "embraces religion as part of culture and deals with religion in the same way as it does with language, race or ethnicity."[63] The Zambian case study will illustrate why this is an appropriate approach to take.

Theoretically, the role of culture is understood as integral to shaping societal identity. Pragmatically, culture is also of great interest to development policy makers because they recognize that culture affects strategy. For example, getting men to wear condoms and women to insist on protected sex in efforts to stop the spread of HIV (pertinent to MDG number 6) is not merely a problem remedied by efficient education, production, and distribution. It involves understanding customs and habits that arise from intangible influences. That is, mobilizing programs to discuss overcoming stigma and making condoms available throughout the "developing" world may create a lot of documentable activities that satisfy funding agencies. But assuming that such programmatic work automatically leads people to use condoms in the manner prescribed and for reasons given by Western public health directives is another matter altogether.

Being cognizant of such discrepancies is critical because it helps make apparent the ways in which questions posed by people experiencing the effects of policy decisions may not be inquiries for which the professional development policy making arena has adequate responses. Why, for instance, should people in impoverished situations expend energy on obtaining and properly using condoms when other needs such as finding food are more pressing? For that matter, why not use nets manufactured to protect sleepers from malaria (also pertinent to MDG number 6) as tools to catch fish when food is scarce and traditional ways of combating the disease, such as draining swamps and burning underbrush, may already exist?

Historically, however, comprehending the power of culture and integrating that knowledge with effective development understanding has been an enormous problem for Western-based initiatives. This is true, in part, because "culture" may be equated with easily discernable phenomena: ethnic food, local arts, traditional dress, peculiar community or family actions, and ritualistic behavior. But such an assessment may not delve into the structures and systems that various societies use to explain and understand the world around them. If in the process of intervening from outside, we have held a focus meeting with the local chiefs; issued a call for proposals throughout a country's NGO network; or made friends with the resident head of the Mother's Union,[64] we may believe that we have adequately dealt with cultural issues because we have attempted to work through those parts of the local environment that already make some sense to us.[65] We may identify the leaders and strategies that appear to correspond to the world as we know it, assuming that relationships and communication channels work similarly everywhere and so proceed with our plans. Meanwhile, we may have given little thought to the possibility that radically different power structures and appropriately effective ways of getting things done exist within other settings but lie beyond our own perceptual boundaries. The danger with this sort of oversight is that genuinely good hearted people and possibly quite beneficial efforts may ultimately meet with frustration and disappointment, because they have been conceived with an insufficient grasp of what culture really entails.[66]

Achieving a sufficient grasp of (or at least gaining an appreciation for) culture's power involves intense self-reflection. Put another way, problems in implementation may arise because we fail to comprehend the cognitive limitations that any culture puts upon its adherents, including our own. Quite simply, we are unaware of our intellectual blind spots. Truman's Point Four Message serves as an example of how this can be the case. The Message certainly could be construed as optimistic and promising but it was hardly introspective. Truman's extensive talk of taking American insights and expertise elsewhere makes the Message's external orientation apparent. Yet, in subsequent years when it became obvious that Point Four-inspired strategies were not bringing American style existence to most of the rest of the world wholesale re-examination of the mission itself did not ensue. Instead, later policy making seemed dominated by an assumption that the way of life Point Four set out to model could not only be replicated but was worthy of propagation and emulation. What appeared to need adjusting were only the methods and theories, not the underlying premises of the entire enterprise.

That healthy self-reflection still remains a stumbling block for many development policy makers and planners may be evidenced in numerous policy-related discussions. For example, talk of designing programs with the objective of *involving* local cultures; using the *help* of particular communities; or of *consulting* interested stakeholders may appear to respect cultural differences and advocate for mutually beneficial interaction.[67] But such talk also suggests something about the ways in which these policies and procedures are being conceived before any plans are ever enacted. This sort of talk suggests, in fact, that certain ideas and values are dominant even if they are simply never mentioned.[68]

There are different ways of accounting for this state of affairs. In a spirit of relentless self-interrogation, we could suggest that Truman's "new program" and everything that the age of development represents have simply been redressed versions of self-interest in domination and control as was true for the colonial era European powers. That is, the post-War Allies never intended to foster equitable peace and prosperity globally; neither does the development industry that Point Four spawned. This is a compelling argument and one that deserves a hearing in any arena where serious discussion of development's intellectual foundations takes place.[69]

But this assessment also presents a predicament particularly for Americans who take their faiths and moral values seriously. For what is to be made of the countless good hearted individuals and inspired organizations that have, frequently at great personal cost, ventured engagement with the wider world because of genuine desires toward service? While it is fair and necessary to examine critically the links between development efforts and efforts at domination and control in general, it would be unfair to suggest that everyone involved in some sort of development effort is consciously intent upon domination and control. Even dubious systemic initiatives may be enacted by people who desire to take the common welfare seriously. Therefore, the possibility of good—doing good or creating environments in which good can occur—exists. As much as a searing critique of Bretton Woods institutional policies is needed, it would be theologically and philosophically irresponsible to assume that every World Bank employee, for instance, is interested in nothing more than global hegemony.[70] Equally irresponsible, of course, would be to surmise that crop failures in India are simply caused by agricultural ignorance or that inter-ethnic violence in sub-Saharan Africa merely documents Africans' inabilities to govern themselves—that Western interests have played no part in such situations.

Rigorous self-reflection, then, is greatly needed on the part of development policy makers and planners but it is a complicated task. Of necessity, scrutinizing assumptions involves reflecting upon cultures and worldviews, specifically the ways in which societies determine what aims and objectives are considered important; what questions are raised about achieving those aims and objectives; and what queries are never even imagined. Only when disparate components of dominant narratives are identified and juxtaposed against possible alternate or hidden narratives can we genuinely assess the effects on others and ourselves.

There is a profoundly moral and ethical dimension to this quandary. For, even if the notion of international development were dispensed with as having come from desires for domination and control, we would still be left with the facts of the day. These are, for instance, that approximately one-third of the world now enjoys a level of comfortable living that the other two-thirds largely does not; that the developed world's ease has come at the expense of others' well-being.[71] While coordinated plans such as the MDGs allow for approaching this inequity from a moral standpoint the responsibility lies heavily upon people of faith, especially in America, to do more. The problem is not just how the age of development was conceived but what needs to happen now that hyper-development is forcing the planet's eco-system out of balance; now that foreign aid and military interventions are increasingly intertwined with wars over natural resources; and now that the halfway point of the MDGs has been passed. Beyond self-reflection, self re-vision is required.[72]

It might seem reasonable to approach this daunting task as a philosophical exercise and there would be some sense in doing so. Certainly development as a product of modern ways of thinking can be rigorously analyzed using reason and empiricism. Such a vantage point would provide room for talking in ethical terms and would fit fairly comfortably with secular viewpoints.[73]

Yet, undertaking such an endeavor with the expectation that it will be sufficient poses two significant problems. The first is that a purely Western philosophical basis of discussion overlooks something crucial about how a big part of the world outside the Global North actually works. This is simply that religion matters deeply elsewhere. In the twenty-first century, much of the world's population is religiously oriented despite twentieth century expectations that modernity and science would make religion outmoded.[74] In fact, while many post-World War II development policy makers and planners tended to assume that religion itself was part of the problem,[75] many others (frequently those who experienced the effects of development policies) have never found legitimate reasons why their worlds have to be split into religious and non-religious realms.[76]

Only within the last couple of decades has there been a substantive turn within the professional development arena toward redressing this intellectual and experiential chasm. Out of Washington, DC, the World Bank has been engaging in formal conversations with faith communities.[77] Likewise, some faith communities have been forthright in expressing the limitations of the Bank's original Bretton Woods-based vision.[78] Pockets of policy makers within the Bank have pressed for greater levels of active partnership with faith communities in areas where Bank projects are being implemented. In Canada, effort has gone toward bridging conceptual gaps inherent in the ways that religious communities think and talk versus the ways that scientifically-based communities think and talk.[79] Recent work from the United Kingdom offers particularly important contributions in this regard, for, as Deneulin asserts "there is no separation between religion and development."[80] So the work of healing the intense divide between "developed" and "developing" worldviews has begun.

But, a second problem inherent in approaching development policy making from a philosophical and scientific-rationalist perspective concerns the possible inhibitions placed on any genuine self-reflection. Innovative rethinking must dare to venture beyond the confines of particular cultural and religious systems or worldviews. For this reason the present work speaks not just of religious but of cosmological understandings in relation to development issues. Effective revisioning will require a broader understanding of "cosmologies."

The Power of Cosmologies to Inform Experience

When religious communities make sense of the universe, or cosmos, they do so by living in and reflecting upon both its commonplace and transcendent elements. When people of faith act, they may do so in the form of rituals, symbols, and customs that merge physical and spiritual realms.

For cultures that do not claim the legacy of the European Enlightenment, full experience of the transcendent and meaningful simply cannot be divided into separate sacred and secular intellectual spaces.[81] Quite rich Native American, Aboriginal Australian, and African scholarship describes what is often termed this type of "indigenous knowledge" and, for the purposes of the present study, attention to African points of view is required. Prominent African intellectuals have long asserted that their cosmological understandings generally describe something a modern Western context would find alien.[82] Certain key points from their work bear on the story to follow.[83] For example, a foundational element in most African cosmologies is the notion that the cosmos itself is unified. Boundaries between what can be considered a spiritual realm and a material one can be malleable while senses of sacredness and mystery pervade everywhere. This state of affairs is partially the case because, second, the world is understood to be populated by a host of active spirits. Ancestors and the recently deceased, for instance, have real import for the presently living and are not revered in memory alone. Third, all of material creation is animated in a tangible manner for which no precise counterpart exists within Western thought. Finally, a certain force accompanies this animation, allowing for supernatural activities such as witchcraft and magic to exercise powerful roles in ways not seen within mainstream Western experience for several hundred years.[84]

Consequently, while "cosmologies" constitute the universes within which people understand themselves to be living, the specific make up of those cosmologies, including what beings populate them and how relationships between spiritual and material realms are understood, can diverge radically across cultures. One cosmology may include elements or beings that another does not. Or a particular cosmology may ascribe power to something or someone that is not the case for another cosmology. In all cases, however, the present work asserts that cosmologies have an impact not only on how people perceive the universe but how they live their daily lives. The story of Zambians trying to make sense of privatization is intended to illustrate one instance in which this was the case.

Narrative as Meaning Making

The disparateness of the above ideas is a primary reason for using a case study method when exploring them. Case studies are increasingly being employed within the academic field of development studies because they allow the transmission of experiential knowledge in ways that theoretical discussions cannot accommodate. Case studies also allow for a narrative approach that has been demonstrably helpful in cross-cultural encounters.[85] A narrative approach acknowledges that we humans constantly story tell in order to understand what is happening around us and why. Without deliberately acknowledging that this is the case we may identify characters with particular roles and responsibilities; chronicle events and actions, giving them beginnings, middles, and endings.[86] We may also recognize that certain themes and ideas exercise force over what people do, over what should or should not happen.

When something such as a dominant development narrative is operating, stories can been seen as discussing identity and self-understanding: who is "developed" and who is not, for example, who is powerful and who is weak. Narrative and story can also reach beyond explanations of group behavior to make life itself intelligible in ways that purely empirical inquiries simply cannot. Narratives such as the story of Zambians experiencing privatization in the early twenty-first century should also be understood as containing legitimate knowledge. This knowledge may lie outside the bounds of statistical charts and measurable objectives but it holds vital information for development policy makers and practitioners. When development policy making and planning are themselves interpreted in storied form, as this work does by articulating a dominant development narrative, such interpretation can further be helpful for illustrating how alternate narratives are often also be present. Disappointments, frustrations, and misunderstandings that are inevitable to cross-cultural engagement are sometimes rendered more comprehensible when it is understood that varying interpretive versions of any situation exist. Depending on that situation, such varying versions can serve both to explain reality and to provide people with coping or survival strategies when necessary.[87]

The narrative study contained here does not pretend to be definitive on issues that are also being studied by other analysts around the world. But it does aim to counterpoint and complement the emerging theoretical literature as well as some of what might be considered standard assumptions within the field of development studies generally (and faith-based involvement specifically). Counterpoint is needed to enrich discussions about what constitute appropriate intervention strategies when the involved parties are radically different. At present, for example, it is often popular to talk of partnership rather than prescription when engaging in development interventions. "Partnership" across socio-economic divides is discussed within multilateral agencies such as the World Bank as well as grassroots communities trying to become involved in the MDGs.[88] As an abstract concept partnership appears to be a means by which some of the mistakes of past patronizing practices could be avoided. But the case to follow illustrates how relationships between different parties can suffer from severe power disparities. This calls into question the genuineness or equitability of partnership as it is generally understood in the policy making arena.

Another objective of this work is to provide complementary details for the growing body of insightful literature that considers development policy in broader fashion than purely economic analyses allow. This literature can be particularly enriched by deeper discussions of moral and ethical issues that are always extant in situations of lived experience. In the present case, for example, Zambian understanding was generally that privatization's progress was going wrong. This was based, at least in part, on moral and ethical norms embedded in cosmologies that saw the universe in particular ways. In such a case, explaining problems with the process by simply using economic, political, or managerial terms would, at best, be inadequate. At worst, it could make no sense at all.[89]

In summary, understanding something of what Zambians often experienced while particular ideas were being enacted in their context can enrich future efforts to devise rules of engagement when disparate parties pursue joint development projects.

Notes

1. Wuthnow, *Boundless Faith*, 235.

2. Theologian George Tinker has presented differing perceptions in a similar manner when discussing disparities between Native American and Euro-American ecological understandings. See, for example, "An American Indian Theological Response to Ecojustice," in *Defending Mother Earth: Native American Perspectives on Environmental Justice*, ed. Jace Weaver (Maryknoll, NY: Orbis Books, 1996), 153-76. Anthropologist Erica Bornstein also uses this sort of terminology in her study of NGO work in Zimbabwe. "Religious beliefs inform the ways that economic development projects are received, interpreted, and accepted in specific social and historical contexts. Religious beliefs also inform the ways in which development projects are constructed by development organizations. They inform the way development is planned, conceptualized, motivated and instituted. These systems of belief, or cosmologies, which link experiences of the material world with conceptions of a spiritual realm (or lack of it) are integral to the processes through which individuals order their worlds and make sense of experience." *The Spirit of Development*, 2.

3. The ethical dimension of industrial societies' interventions into non-industrial societies has been previously noted by other analysts of Africa. See for example, J. Merle Davis, *Modern Industry and the African*, 2nd ed. (London: Frank Cass and Co., Ltd., 1967), 376.

4. Gary T. Gardner, *Inspiring Progress: Religions' Contributions to Sustainable Development* (New York: W. W. Norton and Co., 2006), 14.

5. It appears that compelling theoretical defenses of the concept are increasingly found in the writings of non-Euro-Americans concerning ecological issues. See, for example, what some might term the polemical work of Donald L. Fixico, "The Struggle for our Homes: Indian and White Values and Tribal Lands," in *Defending Mother Earth: Native American Perspectives on Environmental Justice*, ed. Jace Weaver (Maryknoll, NY: Orbis Books, 1996), 29-46. See also Mary Evelyn Tucker and John A Grim, eds., *Worldviews and Ecology: Religion, Philosophy, and the Environment* (Maryknoll, NY: Orbis Books, 1994). I contend that ecological and developmental issues are intimately connected and, therefore, how "worldviews" are discussed and explained in relation to eco-justice issues has relevance for the concept's use in relation to cross-cultural development matters.

6. Stephen C. Pepper, *World Hypotheses: A Study in Evidence* (Berkeley: University of California Press, 1942). See also Steven C. Hayes, Linda J. Hayes, and Hayne W. Reese, "Finding the Philosophical Core: A Review of Stephen C. Pepper's *World Hypotheses: A Study in Evidence*," *Journal of the Experimental Analysis of Behavior* 50, no. 1 (July 1988): 97-111.

7. Hayes, Hayes, and Reese, "Finding the Philosophical Core," 98.

8. That Stephen Pepper inspired Thomas Kuhn's influential work on societal paradigms is not coincidental. Hayes, Hayes, and Reese, "Finding the Philosophical Core," 98. See Thomas S. Kuhn, *The Structure of Scientific Revolutions* (Chicago: University of Chicago Press, 1962). Nor does it seem coincidental that Kuhn's own insights are being routinely cited by a variety of important analysts documenting the current chaotic state of global affairs and raising up a number of alternative visions in response. See, for instance, the work of David Korten, *Getting to the 21st Century: Voluntary Action and the Global Agenda* (West Hartford, CT: Kumarian Press, 1990) and *The Post-Corporate World: Life After Capitalism* (San Francisco and West Hartford, CT: Berrett-Koehler Publishers and Kumarian Press), 1999.

9. Alfred Schutz and Thomas Luckmann, *The Structures of the Life-World*, translated by Richard M. Zaner and H. Tristram Engelhardt, Jr. (Evanston, IL: Northwestern University Press, 1973). Schutz had a good vantage point from which to speak since he was both a social philosopher and an attorney who practiced law for an Austrian banking firm. He is now only rarely cited but the work of his disciples has become classic. See Peter L. Berger and Thomas Luckmann, *The Social Construction of Reality: A Treatise in the Sociology of Knowledge* (New York: Anchor Doubleday, 1966).

10. Alfred Schutz, "On Multiple Realities," in *Collected Papers, vol. 1, The Problem of Social Reality*, ed. Maurice Natanson (The Hague: Martinus Nijoff, 1962), 207-59. See also Berger and Luckmann, *The Social Construction of Reality.* As to how these ideas might affect development policy making and planning Hilhorst has observed that multiple realities and discourses do characterize the experiences of people and organizations involved in development activities. *The Real World of NGOs*, 10.

11. Paul Hiebert, *Transforming Worldviews: An Anthropological Understanding of How People Change* (Grand Rapids, MI: Baker Academic, 2008), 25.

12. This is not to place the entire burden of fault upon professional development personnel or organizations when plans and programs go awry. There are always multiple actors and reasons in such situations. But, from an ethical standpoint, I contend that the burden of responsibility lies with the entities possessing the most power in such cases. When power is understood in economic and political terms as is generally the case in the professional development arena, it overwhelming lies with those who are doing the intervening, that is the propagators of development policies and procedures.

13. See Rist's extended analysis of this term in *The History of Development*, 8-24. Deneulin states that there is "controversy" over what "development" means but, for the purposes of her own introductory work on religion and development, chooses a reference point similar to the one used in this book, namely the immediate post-World War II intellectual and policy making environment. *Religion in Development*, 28-29.

14. See findings produced by the Intergovernmental Panel on Climate Change, "Climate Change 2007: Synthesis Report" (IPCC Plenary XXVII, Valencia, Spain, November 2007), 5, http://www.ipcc.ch/pdf/assessment-report/ar4/syr/ar4_syr_spm.pdf (accessed October 6, 2008).

15. See, for example, arguments made by Thomas Friedman, *Hot, Flat, and Crowded: Why We Need a Green Revolution—and How it Can Renew America* (New York: Farrar, Straus, and Giroux, 2008).

16. According to Haynes, who calls "development" a "vague yet predictive term" this manner of understanding "development" corresponds to theories prevalent during the 1950s and 1960s. *Religion and Development*, 5-7. At the time of this writing, public discussions of ending the wars begun during the presidency of George W. Bush often cite the need for additional development support to provide social infrastructure in Iraq and Afghanistan before peace can prevail. Whether or not peace has ever been the motive behind these or any wars initiated by this country is a much contested question that would take the present work too far afield to address. Ultimately, however, it is a matter for serious reflection in relation to the overall topic of religion and development. Readers interested in a well documented, but disturbing, critique of the history of American foreign policy in this regard can consult Ronald Wright, *What is America? A Short History of the New World Order* (Toronto: Vintage Canada, 2008).

17. See, for example, "New Day, New Way: U.S. Foreign Assistance for the 21st Century," A Proposal from the Modernizing Foreign Assistance Network, 2008, http://www.cgdev.org/content/publications/detail/16210 (accessed October 6, 2008).

18. See http://www.un.org/millenniumgoals/ (accessed October 6, 2008).

19. Deneulin, *Religion in Development*, 48. Haynes notes that at least some prominent individuals within the World Bank have begun to realize the importance of involving faith communities in economic development policy making. *Religion and Development*, 9-10.

20. It should not be assumed, however, that Christian church campaigns for social and economic justice that are formed around or incorporate the MDGs are soundly thought out in theological ways. As Sallie McFague has observed, Western Christianity has too often let the terms of engagement be decided by neoclassical economic theory. "God's Household: Christianity, Economics, and Planetary Living" in *Subverting Greed: Religious Perspectives on the Global Economy*, ed. Paul Knitter and Chandra Muzzafar (Maryknoll, NY: Orbis Books, 2002), 119-36. Bornstein posits further that neoclassical economics and economic development function "as a religious system." She also sees the work of NGOs as "a morally defined form of cultural intervention." *The Spirit of Development*, 2, 169.

21. The United States Agency for International Development (USAID) has made direct connections among democracy, defense, and development in various country strategic plans and, during the second term of the Bush administration, several high ranking officials drew explicit associations between the way the United States formulated development policy and objectives for the President's "War on Terror." Elizabeth C. Parsons, "Provoking the Rocks: A Study of Reality and Meaning on the Zambian Copperbelt" (Ph.D. diss., University of KwaZulu-Natal, 2007), 24-27. On this point see also Wuthnow, *Boundless Faith*, chapter 6.

22. See also Haynes, *Religion and Development*, 207. For some basic statistical analysis of this situation consult Jonathan Bonk, *Missions and Money: Affluence as a Missionary Problem . . . Revisited* (Maryknoll, NY: Orbis Books, 2006), 3-10 and Joseph E. Stiglitz, "Globalism's Discontents," *The American Prospect*, Winter 2002, A16-21.

23. The first MDG lists as its first target, between 1990 and 2015, halving the proportion of persons living on less than $1/day. In 2008, the World Bank reported that more people than previously estimated fell into this category (about 1.4 billion). http://www.un.org/millenniumgoals/2008highlevel/pdf/newsroom/Goal%201%20FINAL.pdf (accessed October 6, 2008).

24. See, for example, the efforts of Episcopalians for Global Reconciliation, http://www.e4gr.org/mdgs/fast_facts.html (accessed March 25, 2009) which addresses the entire MDG package and those of the Muslim organization, United for Change that focuses on malaria eradication (MDG #6), http://www.muslimsunitedforchange.com/ (accessed March 25, 2009).

25. Deneulin, *Religion in Development*, 29.

26. Haynes, *Religion and Development*, 7; Deneulin, *Religion in Development*, 28.

27. In addition to Bornstein's work see that of Hilhorst, *The Real World of NGOs* and Occhipinti, *Acting on Faith*.

28. Deneulin, *Religion in Development*, 40; Haynes, *Religion and Development*, 1.

29. There is also a risk that, in attempting to avoid a sense of hegemony or essentialism that can accompany discussions of worldviews and group identities, inadvertent negation of another's viewpoint can take place. That is, when members of one group experience themselves to be living within a context that operates according to values and systems that are foreign to them, and that context is described by the group as being mainstream or dominant over against their own understanding, this should be taken seriously even by outside observers who are sensitive to the perils of essentialism. Fixico, for instance, speaks unhesitatingly of a "mainstream culture in America" into which "minority groups" try desperately to integrate. "The Struggle for Our Homes," 31.

30. This terminology should be taken in a spirit compatible with Deneulin's observations about the evolving history of development thought. *Religion in Development*, 28-51. She also uses the term "dominant script" when discussing development theory's treatment of religion. *Religion in Development*, 6. A "dominant development narrative" is one way of looking at particular strains of thought at a particular point within a shifting intellectual and experiential context. When this case study refers to a dominant development narrative, it is in reference to general, or broad, outlines of a depiction or interpretation of "development" that is detectable in official policy documents and arenas of public discussion. Similarly, Occhipinti identifies what she terms a dominant development "discourse" that is "not like a cage, a restrictive and all-encompassing framework of fixed ideas and meaning." Rather, it can be understood as a dynamic process involving, ideas, resources, rules, and the use of power that is "promoted by particular networks of agents, within which economic, political, and social change occur." *Acting on Faith*, 45. What is especially noteworthy about her analysis in terms of the present study is the further observation that developmental discourse approaches lived circumstances from the standpoint of "problem" assessment. The intellectual starting point for development work generally involves an assumption that the arenas in which the work is to unfold contain certain problems. How these problems are defined determines the strategy that development agencies and projects will then enact.

31. In writing about Native American eco-justice issues, Tinker supports a storied approach as being the most powerful means for resolving destructive disparities between worldviews and devising theologies appropriate to the current era. "Theology must become an exercise in expressing the self-identity of whole communities. For this sort of theology we need stories rather than treatises, essentialist discourse, problem-resolution, or structuralist puzzle-solving." "An American Indian Theological Response to Ecojustice," 166. The present work aims to employ a similar spirit of thought.

32. Marshall and Van Saanen, *Development and Faith*, 2. See also the introduction and first chapter of Haynes, *Religion and Development*.

33. Haynes, *Religion and Development*, 2.

34. In Zambia's case, the activities of the various chartered companies set up by the colonial powers would fall into this category. Additionally, work inspired by David Livingstone, who is discussed briefly in chapter three, may be thought of—at least to some degree—in this vein. This is not said to conflate the very different motivations of entrepreneurial and missionary groups that often operated in the same regions. The point is, rather, that in Western understanding physical construction and the institution of material infrastructure have often been considered, if not completely means to ends, at least crucial to an overall objective reaching beyond the material world itself. This has not always been the case, however, as Hovland points out in her study of the Norwegian Mission Society. In that setting, material or physical goods were not considered "bait" for religious conversion. In fact, to do so, would pose profound theological difficulties. Ingie Hovland, "Who's Afraid of Religion? Tension Between 'Mission' and 'Development' in the Norwegian Mission Society," in *Development, Civil Society and Faith-Based Organizations: Bridging the Sacred and the Secular*, ed. Gerard Clarke and Michael Jennings (Basingstoke: Palgrave Macmillan, 2008), 171-86.

35. Dana Robert, *Christian Mission: How Christianity Became a World Religion* (Malden, MA: Wiley-Blackwell, 2009), 5; Gustavo Esteva, "Development" in *The Development Dictionary: A Guide to Knowledge as Power*, ed. Wolfgang Sachs (London: Zed Books, 1992), 6-12; Chinweizu, *Decolonizing the African Mind* (Lagos: Pero Press, 1987). Wuthnow's sociological study of American Christian international activity also raises significant questions about the contemporary situation. See *Boundless Faith*.

36. Haynes asserts that for modernity to take place, religion had essentially to get out of the way. *Religion and Development*, 6. See also Deneulin, *Religion in Development*, 52.

37. Paul Hiebert links the materialist emphasis of empirical scientific investigation to the socially constructed consumerist emphasis of modern society. He posits that science's necessary focus on physical evidence eventually helped lead to more popular collective rejection of spiritual and supernatural realities. "Concern for this life rather than eternity led to a stress on physical comfort and material abundance." *Transforming Worldviews*, 148. Dualism inspired by an understanding of the world as divided into sacred and secular realms ultimately affected Christian mission work (endeavors themselves long understood as being precursors to modern day development initiatives) resulting in yet more divisions. "Missions" came to be thought about in terms of evangelism, or proclamation seeking converts, and Social Gospel, or service. Hiebert, *Transforming Worldviews*, 154. In considering why religion has been so neglected in development policy making, Gerard Clarke uses the term "materialistic determinism." This, he says, accounts for "the neglect of nonmaterial, especially religious, motivations in explaining individual and institutional behavior" and is integrally involved in the secularization thesis out of which Western development policy making has arisen. Gerard Clarke and Michael Jennings, eds., *Development, Civil Society and Faith-Based Organizations: Bridging the Sacred and the Secular* (Basingstoke: Palgrave Macmillan, 2008), 17ff. See also Deneulin, *Religion in Development*, 52ff.

38. Bonk, *Missions and Money*, 20. See also Haynes who uses the word "emulate" in his argument on this point. *Religion and Development*, 6.

39. See, for example, Ogbu U. Kalu, *Power, Poverty, and Prayer: The Challenges of Poverty and Pluralism in African Christianity, 1960-1996* (Trenton, NJ: Africa World Press, 2006), 75-100; Wuthnow, *Boundless Faith*; and Dambisa Moyo, *Dead Aid: Why Aid is Not Working and How There is Another Way for Africa* (London: Allen Lane, 2009) contain multiple references to inappropriate interventions. See also another study concentrating on the mining industry in Africa. John Hollaway, *All Poor Together: The African Tragedy and Beyond* (Johannesburg: Capricorn Books, 2000).

40. Haynes contends that, as the concept of development began to evolve in the post-World War II world, the problems of so-called developing countries were initially defined as technical issues that could be resolved with proper training and management. *Religion and Development*, 6.

41. Chinweizu, *Decolonizing the African Mind*, 18. The sustainable development movement has offered a harsh critique of this sort of practice, noting particularly the cruel irony of the United States telling other countries to do what it has done materially when this country's own material existence has now far exceeded Earth's capacity. See Wright, *What is America?* 192ff.

42. As of this writing, the majority of sources in the burgeoning field of religion and development generally and those that particularly critique American-based policies and procedures originate from outside this country as the citations in the present work should make clear.

43. Esteva, "Development," 8; Frantz Fanon, *The Wretched of the Earth*, 1st Evergreen Black Cat ed. (New York: Grove/Atlantic, Inc, 1968); Karen Fields, *Revival and Rebellion in Colonial Central Africa* (Princeton: Princeton University Press, 1985).

44. Sachs, *The Development Dictionary*, 2.

45. Deneulin's introductory survey notes the importance of certain historical events in post-World War II America but she also uses other means of approaching the intellectual history of development. For example, she highlights a series of ideas such as moder-

nization, basic needs, and human rights that have variously played important roles. *Religion in Development*, 29. Not all of this material originated within the United States. I have chosen a more historical route deliberately focused on the role of the United States as a way of exploring how certain taken for granted notions within our borders do not appear so from the outside.

46. Henry Veltmeyer asserts that the Bretton Woods system was specifically designed to ensure the existence of a capitalist system in which the United States would dominate. *Illusion or Opportunity: Civil Society and the Quest for Social Change* (Halifax: Fernwood Publishing, 2007), 6-7.

47. See Rist, *The History of Development* and Sachs, *The Development Dictionary* for extended discussions of this subject.

48. Rist, *The History of Development*, 72-73.

49. Haynes also notes that Truman's agenda was, in part, to "bind [poor countries] ideologically and financially to the United States at the beginning of the Cold War." *Religion and Development*, 2.

50. Rist, *The History of Development*, 70ff.

51. See also Haynes, *Religion and Development*, 3.

52. See Hilhorst, *The Real World of NGOs* and Occhipinti, *Acting on Faith* for case examples of this type. From a Native American perspective, Tinker asserts that the nation-state has philosophical, as well as political and economic roots. This "artificial" structure owes much to the Western propensity for individualistic thinking and systemically erodes not only indigenous sovereignty but communal values on the whole. "An American Indian Theological Response to Ecojustice," 165-73.

53. See, for example, Kalu, *Power, Poverty, and Prayer*, 75; Korten, *Getting to the 21st Century*, 2, 5-7; David Bosch, *Transforming Mission: Paradigm Shifts in Theology of Mission* (Maryknoll, NY: Orbis Books, 1991), 356.

54. This is the premise of such works as Korten, *Getting to the 21st Century* and especially Amartya Sen, *Development as Freedom* (New York: Alfred A. Knopf, 1999).

55. Wuthnow draws a straight line from the twentieth century phenomenon of nongovernmental organizations to faith-based missionary efforts of the nineteenth century. *Boundless Faith*, 118. See also the definitional discussion of faith-based organizations and their origins in Elizabeth Ferris, "Faith-Based and Secular Human Organizations," *International Review of the Red Cross* 87, no. 1 (June 2005): 315.

56. Korten, *Getting to the 21st Century*, 2, 67. See also excellent summaries of various definitions and perspectives provided by Bryant Myers in *Walking with the Poor: Principles and Practices of Transformational Development* (Maryknoll, NY: Orbis Books, 1999), 94-110.

57. Haynes, *Religion and Development*, 3-4, 68; Deneulin, *Religion in Development*, 29-30.

58. Sen, *Development as Freedom*, 3.

59. Haynes, *Religion and Development*, 14.

60. Deneulin, *Religion in Development*, 60. See also Tinker's discussion of culture, language, categorical labeling, and difficulties posed on non-Western ways of understanding. "An American Indian Theological Response to Ecojustice," 155.

61. Haynes, *Religion and Development*, 14. Further, Stephen Ellis and Gerrie Ter Haar discuss "religion" in Africa as encompassing magic and other experiences of the supernatural. Stephen Ellis and Gerrie Ter Haar, *Worlds of Power: Religious Thought and Political Practice in Africa* (London: Hurst and Co., 2004), 3.

62. Haynes, *Religion and Development*, 14-15.

63. Deneulin, *Religion in Development*, 41. She goes on to say that only processes of development that allow people to live according to the dictates of their own religious and cultural communities should be considered legitimate. Both Deneulin and Haynes recognize inherent difficulties in defining "religion." Deneulin, who contends that the manner in which "religion is defined has important consequences for its treatment in development thinking and practice," opts for a definition of religion as "a tradition" that "rests on the fundamental agreement that a good life is one which abides by God's commands." *Religion in Development*, 63, 70. In Haynes's mind, "religion" reaches beyond the material realm into a spiritual one that can include the supernatural. He also recognizes religion as playing both constructive and destructive roles in development. *Religion and Development*, 6, 14-16.

64. Mother's Unions are powerful networks of women leaders frequently found in mission churches throughout the world. They originated through the work of an Anglican woman, Mary Sumner, in the 1800s.

65. Robin Horton makes a similar observation when discussing differences between Western and African thought systems. In analyzing the difficulties of translating local African languages into one of the colonial languages (English, French, or German for example), he admits that the process involves putting African concepts into Western languages and thought patterns. This is a very imperfect activity since one must search for "the appropriate Western conceptual pigeon-holes for African concepts and thought patterns." Sometimes, he says, Western concepts must be stretched. Robin Horton, *Patterns of Thought in Africa and the West: Selected Theoretical Papers in Magic, Religion, and Science* (Cambridge, UK: Cambridge University Press, 1993), 303. The point here is that when working cross culturally we may unwittingly, and perhaps poorly, fit concepts and actions from the other culture into a frame of reference that makes sense from our own cultural standpoint.

66. Clifford Geertz uses the term "thick description" to capture the idea of well nuanced, cross cultural engagement and analysis. Clifford Geertz, *The Interpretation of Cultures* (New York: Basic Books, 1973).

67. A few such examples include the following: "International Development Association Program Document for a Proposed Credit in the Amount of SDR 27.4 Million (US$40 Million Equivalent) to the Republic of Zambia for an Economic and Growth Management Credit," Report 29294 ZA (Washington, DC: The World Bank, 2004), 37; Oliver Campbell White and Anita Bhatia, *Privatization in Africa* (Washington, DC: The World Bank, 1998), 117.

68. The description for a master's degree in International Educational Development at Boston University describes its mandate thusly, "Our graduates are applying their skills and knowledge around the globe and here at home. Some work with the World Bank, UNICEF, and UNESCO (both at the mission levels and at the headquarters) to improve project planning and evaluation within those multilateral organizations. Others work with consulting development organizations such as World Education, Education Development Center, and the Academy for Educational Development . . . Today's world includes a large population that is deprived of access to democracy or social justice, and loyalty is exclusively to the family or ethnic group. Ignorance, hunger, poverty, and disease affect more people today than they did fifty years ago. We have a lot of work to do, and we need students who are committed to making a difference in the world." http://www.bu.edu/sed/programsinternational.htm (accessed December 29, 2009). The Center for International Development at Harvard University characterizes its mission in the following manner, "The CID's primary activities and programs seek to: Change the way that growth strategies are conceived, designed, and implemented; Reinvent produc-

tion-related policies to facilitate countries' move to higher productivity activities; Extend markets to the underserved and empower the disenfranchised; Improve service delivery in education, health, and other social services; Design institutions, policies and practices that promote sustainable development that meets human needs while conserving the earth's life support systems." http://www.cid.harvard.edu/cidinformation/ (accessed December 29, 2009). Both of these cases contain conceptual assumptions that many non-Westerners could legitimately challenge from within their own philosophical and theological worldviews.

69. See, for example, Wright, *What is America?* 192.

70. See remarks on altruism in Robert Chambers, *Whose Reality Counts? Putting the Last First* (London: Intermediate Technology Publications, 1997), 13.

71. Susan George, "How the Poor Develop the Rich," in *The Post-Development Reader*, eds. Majid Rahnema and Victoria Bawtree (London: Zed Books, 1997), 207-13.

72. The need to re-vision and fundamentally change our thinking habits is increasingly being talked about in some creative, inspiring ways relative to the climate crisis. Eco-theologian Larry Rasmussen, who asserts that dominant sustainable development and "green" technology global models of understanding are inadequate for the task ahead, maintains that the earth's "distress is a crisis of culture." Larry L. Rasmussen, *Earth Community Earth Ethics* (Maryknoll, NY: Orbis Books, 1996), 7. Adequately meeting the challenges of climate change, pollution, and resource wars will, he says, require abandonment of certain habits of mind that tragically separate humanity from the rest of creation. This he sees as inevitably tied to religious experience in a broad sense. Rasmussen, *Earth Community Earth Ethics*, 31-34. David Korten, arguing from the standpoint of an economist and international development expert, agrees that what is called for at the most foundational level is a spiritual rather than political or economic matter. Korten, *The Post-Corporate World*, 281. This argument appears toward the end of the cited work and is developed more fully in his subsequent writings. Many Americans passionately agree with these sentiments and there has, of late, been movement particularly on the part of evangelical Christian Americans to meld political action with religious conviction on behalf of the climate crisis. The Rasmussen and Korten viewpoints remain, however and unfortunately, minority opinions within general American-based discussions of development policy making. Further, the work of evangelical Christians does not, as yet, appear to have holistically joined thinking about the climate crisis to analysis of systemic issues behind hyper-development.

73. See Rist, *The History of Development*, for an excellent example of this approach.

74. Karen Armstrong, *A History of God: The 4,000 Year Quest of Judaism, Christianity and Islam* (New York: Ballatine Books, 1993).

75. Haynes, *Religion and Development*, 1.

76. It should not be inferred from this discussion of Western philosophy and scientific-rationalism that non-Western ways of thinking are necessarily irrational. In fact, there is a long tradition of excellent scholarship detailing the rationality of various African thought systems. See Horton, *Patterns of Thought in Africa and the West*. See also Edward E. Evans-Pritchard, *Witchcraft, Oracles and Magic Among the Azande* (Oxford: Clarendon Press, 1937) and Kwame Anthony Appiah, *In My Father's House: Africa in the Philosophy of Culture* (Oxford: Oxford University Press, 1993). This issue will further be discussed in relation to the Zambian mining case in a later chapter.

77. Wendy Tyndale, "The World Faiths Development Dialogue: An Interfaith Response to Poverty and Development in an Age of Globalization," in *Development Matters: Christian Perspectives on Globalization*, ed. Charles Reed (London: Church House Publishing, 2001), 106-15; Marshall and Van Saanen, *Development and Faith*, 195ff.

78. Marshall and Van Saanen, *Development and Faith*, 195ff.

79. Sharon Harper ed., *The Lab, the Temple, and the Market: Reflections at the Intersection of Science, Religion, and Development* (Bloomfield, CT: Kumarian Press, 2000).

80. Deneulin, *Religion in Development*, 4. Similarly, Bornstein in writing about Zimbabwean NGO and FBO development efforts asserts, "In African Christian faith, the realms of the spiritual and material cannot easily be separated: development is both spiritual and material." *The Spirit of Development*, 49.

81. As Bornstein has observed, "The degree to which one can assert that an economic context is 'secular' in southern Africa (or anywhere) may be governed largely by Western assumptions of divisions between the secular and the religious." *The Spirit of Development*, 3.

82. While much identification of African cosmologies was first formally codified by European missionaries and scholars there is also an ample and growing corpus of ground breaking African work. Some prominent African scholars who have written on this subject include John Mbiti, *African Religions and Philosophy* (London: Heinemann, 1969) and Kwame Bediako, *Theology and Identity: The Impact of Culture Upon Christian Thought in the Second Century and in Modern Africa* (Oxford: Regnum Books, 1992). See also, Kalu, *Power, Poverty, and Prayer*.

83. This portion of the discussion draws heavily upon Bediako's interpretation of Harold Turner's influential work. Kwame Bediako, *Jesus in Africa: The Christian Gospel in African History and Experience* (Akropong-Akupem, Ghana: Regnum Africa, 2000), 87-88.

84. For further reading on African cosmologies in addition to Kalu, *Power, Poverty and Prayer* and Horton, *Patterns of Thought in Africa and the West*, consult any of the following texts: Kwesi Dickson and Paul Ellingworth, *Biblical Revelation and African Beliefs* (London: Lutterworth Press, 1969), 137-58; Ellis and Ter Haar, *Worlds of Power*; Evans-Pritchard, *Witchcraft, Oracles and Magic Among the Azande*; Peter Geschière, *The Modernity of Witchcraft: Politics and the Occult in Postcolonial Africa* (Charlottesville, VA: University Press of Virginia, 1997); Max Gluckman, *Custom and Conflict in Africa* (Oxford: Basil Blackwell, 1963), 81-108; Barry Hallen and J. O. Sodipo, *Knowledge, Belief and Witchcraft: Analytic Experiments in African Philosophy* (London: Ethnographica, 1986); Mbiti, *African Religions and Philosophy*; Henrietta L. Moore and Todd Sanders, *Magical Interpretations, Material Realities: Modernity, Witchcraft, and the Occult in Postcolonial Africa* (London: Routledge, 2001); Sipho Mtetwa, "African Spirituality in the Context of Modernity," *Bulletin for Contextual Theology* 3, no. 2 (1996), http://www.sorat.ukzn.ac.za/theology/bct/tonyb.htm (accessed March 23, 2009); Placide Tempels, *Bantu Philosophy* (Paris: Présense Africaine, 1959); Nimi Wariboko, *The Depth and Destiny of Work: An African Theological Interpretation* (Trenton: Africa World Press, 2008).

85. Anthony O. Balcomb, "The Power of Narrative: Constituting Reality Through Storytelling," in *Orality, Memory and the Past: Listening to the Voices of Black Clergy Under Apartheid*, ed. Philippe Denis (Pietermaritzburg: Cluster Publications, 2000), 49-62. Langdon Gilkey, *Nature, Reality, and the Sacred: The Nexus of Science and Religion*, Theology and the Sciences Series (Minneapolis: Fortress Press, 1993), 86; Haynes, *Religion and Development*, 56; Tinker, "An American Indian Theological Response to Ecojustice," 166. See also Ellis and Ter Haar who make ample use of stories and rumors in their work on African religions and policies asserting that these forms of communication are especially important and credible in the orally based societies of Africa. They go on to link such narratives directly to increasing understanding of how Africans make sense

of their worlds. "Studying the rumours and popular stories that abound in Africa also provides us with an outstanding means for investigating African epistemologies, in other words the theory of the methods or grounds of knowledge. Whether or not a particular rumour is true is not of immediate relevance for our purposes: we wish to demonstrate how rumours provide insights into the way in which people apprehend ideas." *Worlds of Power*, 7. Similarly, Luise White, in her anthropological and historical investigation of vampire stories that circulated during the colonial era posits that "the line between fact and story, indistinct as it is for most historians, is made concrete by people who believe that some agency of the colonial state captured Africans and took their blood." Luise White, *Speaking with Vampires: Rumor and History in Colonial Africa* (Berkeley: University of California Press, 2000), 308. As to the credibility of narrative generally, she observes, "Stories and rumor are produced in the cultural conflicts of local life; they mark ways to talk about the conflicts and contradictions that gave them meaning and power." *Speaking with Vampires*, 312.

86. Balcomb, "The Power of Narrative," 51.

87. James Scott, *Weapons of the Weak: Everyday Forms of Peasant Resistance* (New Haven: Yale University Press, 1985), 29.

88. Marshall and Van Saanen, *Development and Faith*; Andrew J. Kirk, *What is Mission? Theological Explorations* (London: Darton, Longman, and Todd, 1999). See also Wuthnow on the popularity and limitations of the partnership concept in reference to American Christian global engagement. *Boundless Faith*, 244-46.

89. Tinker uses similar language when discussing how the intellectual context of a particular World Council of Churches study document generally appeared to Native Americans who were asked to participate in the study process. "An American Indian Theological Response to Ecojustice," 155.

♀

Chapter Two

Research Methods

One of the biggest challenges of qualitative cross-cultural research is making sure that approaches taken are appropriate for obtaining the information sought. When the researcher is a Westerner working in Africa, this is particularly difficult. The reasons for this are many and varieties of obstacles have been encountered by past Copperbelt researchers.[1] Because this work argues for thoughtful, cross-cultural engagement, it follows that methods matter. So this chapter offers extended documentation particularly on how the field work was conducted.

A perennial problem of much development-related research is a tendency to conduct the exercise as efficiently as possible on the part of those who are gathering the data.[2] This often means engaging in short term sight visits and interviewing the locals who are easiest to find. Externally based researchers may fly into an area for periods of only weeks or months having simply read about the place before arriving. When working with specific on the ground organizations, researchers may approach mid- and upper-level managers residing in urban areas because complicated travel logistics make rural site visits more difficult to arrange. The situation grows more complex when linguistic boundaries exist. It is easiest for outsiders working under time constraints to find interviewees who are fluent in a common language than it is to become fluent in the local language or to arrange intricate systems of trusted translators and language checkers needed to come close to a product representing candor on the part of their informants.[3] Under such circumstances of efficiency data is certainly produced.[4] But the extent to which that data accurately reflects the experiences of ordinary people living with the consequences of development policies is questionable.

Despite limitations inevitable to any study of this type, at least an effort was made here to do things differently. The following pages discuss methods used to gather information and to identify procedural difficulties or challenges posed by the research questions as well as strategies employed to address them.

Even though it did not constitute part of the fieldwork, the time I spent as a teacher/trainer in Zimbabwe and Zambia during the early 2000s helped me gain familiarity with the area in which the case study unfolded. Experience living and working in Africa provided historical and cultural information that could never have come from books. During those years, I also learned much about how to interact with Africans; what to look for when trying to determine people's priorities and what to question; the types of situations and people to trust and those about which to be skeptical. Living and working with Zimbabweans and Zambians of both African and European heritages blessed me with a small circle of friends and colleagues who provided insights and contacts that would have been impossible to find had I simply come to the continent for this study alone. Time spent on the Copperbelt also allowed me to absorb much about the country and province that has been integral to this study's context and that gave the research a head start.

Following completion of full time residency in Southern Africa, I returned six times to the continent to conduct fieldwork for this study. Between the latter half of 2003 through the end of 2005, I spent a combined total of approximately eight months on site and, in the interim, kept in contact with a network of study participants in Africa via email and telephone.[5] Two Zambian research assistants also maintained contacts and gathered information when I was not present; more will be said about them below. From my home in the Boston area, I was able to conduct background reading; monitor international news via the Internet; and interview other study participants in the United States and Canada.

Field Data Collection

While secondary sources helped shape and buttress this study, work with primary sources and materials constituted the bulk of activities conducted in the field.

A total of 212 persons contributed to this research project and Table 2.1 records basic demographic data about them. Since qualitative research strives for types of knowledge that are simply unattainable through quantitative methods, emphasis goes toward depth of engagement with sources. This contrasts with, for example, statistical surveys that may go out to vast numbers of informants but not necessarily engage with those informants over extended periods of time. In the case of this study, many times we had repeated interactions with participants even though the statistical chart records only the total number of informants who participated. By far, the majority of participants in this study belonged to the Zambian working class—miners, their family members, or friends.[6]

A number of expatriates participated as well, although this group should not be viewed as a direct counterpart or opposite to any Zambian group. Primarily for historical reasons, the study concentrated on experiences of European and North American expatriates. Although, as the next chapter will discuss, there has been Arab and Indian influence on the African continent for centuries and, in the last decade, a dramatic increase in Chinese influence as well, the particular relationship between the European and African continents brought on by colonialism was of special interest to the topic at hand.

There were also logistical constraints that inhibited data collection from non-European mining expatriates. At the outset of the field work period, the only mine that had been sold to Indian investors during the privatization process

(Roan Antelope) had already closed. When it reopened about two years later, it did so under European ownership.[7] The only other mine to be sold to Indian financiers during the period covered by this investigation was Konkola Copper Mines. Despite concerted attempts, I was never permitted access to expatriates or executives of this company. Chinese investment in and oversight of the mines was comparatively limited at the time of this study. The one set of companies (NFC Africa Mining and the BGRIMM explosives factory) near Chambishi that did operate under Chinese control in the early 2000s is the subject of consideration from a Zambian perspective in subsequent chapters of this work. Chinese expatriate staff members were seen as notoriously aloof from the local communities in which they resided and did not appear to demonstrate significant interest in the organizational infrastructure that accompanied the mining industry. NFC was not, for example, a member of the Chamber of Mines at the time.[8] After completion of the formal fieldwork period, informal conversations that continued with expatriate contacts revealed a rapid influx of Chinese workers into the Zambian mining industry and the country as a whole. Such anecdotal reports accord with more formal observations about the speed and intensity of China's involvement with the continent.[9]

Information Gathering Processes with Zambians

Contacts with local residents of the wider Copperbelt community arose from friendships established during the time I lived in Southern Africa. Since my initial assignments had involved working with Zambians via the Anglican Church, I actually had closer relationships with Zambians than with expatriates going into the research process. These contacts served as the starting point for entrée into the local mining communities. I was also fortunate to maintain formal affiliation as a research consultant with Mindolo Ecumenical Foundation (MEF) in Kitwe where I had previously taught. Mindolo is a respected training institution in Zambia that has roots in the African mining community as will be seen in successive chapters. Its connection gave the study credibility and assured members of the Zambian mining community that I was not affiliated with corporate management.[10]

This association did not, however, thoroughly allay local people's concerns about the investigation. According to my research assistants and other trusted Zambian colleagues, some suspicions arose simply from the nature of the inquiry itself. Since the study took seriously ideas and concepts not normally associated with professional development initiatives, there were occasions when participants appeared truly puzzled—and therefore hesitant—as to why we were interested in the subjects we discussed.[11] Toward the end of the fieldwork, one of my research assistants said it was as if we had gone into people's homes and they had offered us the best of whatever they had there but we said, "No. We are interested in the trash you have thrown out back."

Table 2.1. Demographic Statistics of Study Participants

Data type	**Number**
Total number participants	212
Total number men	164
Total number women	48
Total number age 18-29	37
Total number age 30-49	123
Total number age 50+	52
Total number African	169
Total number Asian	1
Total number European/North American	42
Men age 18-29	28
Men age 30-49	94
Men age 50+	42
African men	130
Asian men	1
European/North American men	33
African men age 18-29	27
African men age 30-49	77
African men age 50+	26
Asian men age 18-29	0
Asian men age 30-49	1
Asian men age 50+	0
European/North American men age 18-29	1
European/North American men age 30-49	16
European/North American men age 50+	16
Women age 18-29	9
Women age 30-49	29
Women age 50+	10
African women	39
Asian women	0
European/North American women	9
African women age 18-29	9
African women age 30-49	26
African women age 50+	4
Asian women age 18-29	0
Asian women age 30-49	0
Asian women age 50+	0
European/North American women age 18-29	0
European/North American women age 30-49	3
European/North American women age 50+	6

Table 2.2. Representative Listing of Study Participants' Job Types and/or Societal Roles

Acid off loader	Fitter	Pipe layer
Accountant	Folklorist	Plumber
Administrator	Former mine electrician's wife	Police officer
Air loader machine repairer	Former mine contract worker	Policy analyst, economist
Ambulance driver	Former miner's wife; daughter	Past President of Zambia
Anthropologist	Futures/commodities trader	Priest
Archival assistant	Geologist	Projects manager
Artisan	IT manager	Pump operator
Assistant engineer's wife	IT systems staff member	Refinery worker
Blaster	Safety and health officer	Rigger
Boiler pump supervisor	Security services manager	School inspector
Business owner/contractor	Heavy equipment operator	Secretary
Butcher	Hotelier	Section boss
Casual worker	Human Resources director	Security guard
Chamber of Mines worker	Development analyst	Senior diplomat
Check point clerk	Jackhammer operator	Shaft mine captain
Chief Executive Officer	Locomotive driver	Shaft worker's wife
Chief Financial Officer	Lodge owner	Shift boss
Chief Mining Officer	Machinist	Shoe repairman
Chief Services Officer	Manager (ZCCM)	Shop owner
Civil servant	Mechanic	Smelter expert
Contracting company manager	Mechanical fitter	Smelter manager
Contracting company owner	Mechanical fitter, smelter	Surface fitter
Cultural analyst	Member, diplomatic corps	Taxi driver
Database administrator	Metallurgist	Teacher
Draftsman	Mine owner's wife	Technical consultant
Driller	Miner's brother	Timber company owner
Drilling contractor	Miner's daughter	Timberman
Driver, contracting company	Miner's sister	Traditional healer
Economic advisor	Miner's son	Travel agent
Electrician	Miner's wife	Underground manager
Emerald mine owner	Newspaper reporter	Unemployed resident
Engineer	Nurse	Vegetable seller
Entrepreneur	Office manager	Winch operator
Explosives instructor	Physician	Winding engineer driver
Farm owner	Physician's wife	World Bank contractor

Although the theoretical parameters of the study were clear, there were numerous practical challenges to overcome when deciding where and with whom to begin. I already knew something of the pressure many Africans could feel to say what they think Westerners wanted to hear.[12] I also knew how easily I could accidentally stifle cross-cultural conversation by speaking too quickly, posing closed rather than open-ended questions, or generally being more direct and assertive than would be customary in non-American settings. Additionally, I was familiar with the teaching techniques of Paulo Freire that emphasize starting from what people know and building upon that.[13] If I picked out the problems I saw, I might not identify those that Zambians saw.[14] So it seemed imperative to begin at a point of their choosing rather than my own. I had further been advised that Zambians were "allergic" to paper questionnaires and surveys because these connoted past government efforts at control and monitoring.[15] Whatever was done would have to be conducted orally, in the vernacular, and informally.

Deciding that simplicity would be best we settled on having Enock Muthwejile, my principal research assistant, do a small random survey within the mining compounds. This survey consisted of two questions that might be thought of as sensory and cerebral, respectively. Question 1, "Regarding the mines, what do you see happening around you?" was designed to elicit people's intuitive knowledge. Question 2, "Why do you think these things are happening?" was designed to engage with rational knowledge or logic. We also conducted these brief interviews in disparate locations as a way of gauging the broadness or narrowness of concerns expressed and to guard against respondents conferring with each other before participating. As a further safeguard against overly influencing people's responses, I did not share with my research assistant in great detail my own ideas or specific points of focus for the study. He knew broad generalities only. We originally intended to take a large sample: begin with 50 participants and see how wide ranging or how uniform their answers were, then make a decision about where to go from there. After 20 interviews, Enock remarked that it was as if everyone lived in the same house. The uniformity of responses was notable but not surprising in one sense given the Copperbelt's social history.[16] And, as the study unfolded, we received repeated corroboration of findings first expressed in this little situational analysis survey.

Identifying Local Contacts

There were at least three reasons why I wanted to have a large and disparate group of local contacts or informants for the entire study. As with the situational analysis, the first reason was as a double check on the information gathered. If we had informants in various Copperbelt towns, the chances of our obtaining candid insights would be higher because people would be less able to coordinate their stories. Second, spreading out the study would be another way of dissipating concerns as to what we were really up to. By saying the study was of the Copperbelt situation and not specifically of one mining company we were able to reduce possible feelings of intrusion or threat. Third, taking in a larger swath of the Copperbelt would allow us to explore information or situations about which I was already aware and curious. These included the complex relationship between Nchanga/Chingola and the Anglo American group of companies; the history of the Roan Antelope Mine in Luanshya; and the 1970s era mine disaster in Mufulira. We began looking for local contacts in towns throughout the Copperbelt and ultimately found our most consistent ones in Chingola, Kitwe, Luanshya, and Mufulira. Brief biographical sketches of our main Zambian informants as well as how we found and worked with them follow below.

Most interactions with our Zambian contacts were quite informal and we spent considerable time in people's homes, at church functions, and in the townships including their marketplaces and bars. Much of the best information came through conversations not conducted under obvious or heavy Western influences. This meant emphasizing the use of Copperbelt Bemba[17] rather than English and, upon occasion, sending the research assistants into settings where it simply would not have been appropriate for a foreigner to go. My presence was inevitably obtrusive especially at first but rarely unpleasantly so. Over time and after repeated visits, it always seemed a success when those we talked with began using more Bemba than English and addressing each other or the research assistants more directly rather than the principal researcher.

The ambivalent influence of Western Christian connections also deserves mention here. Christianity provided a common starting point in that we were able to move with credibility into the townships using Anglican Church networks rather than mining company or union networks. In a testimony to the power of Christian churches within Zambian society interviews with former Zambian Presidents Kenneth Kaunda and Frederick Chiluba were obtained through the Zambian Anglican Council. Acknowledgment of Christianity's presence could also appear threatening, however. For example, there were a few times when overt references to Christianity obviously changed the dynamics of conversations. I will never know the extent to which this affected the study but my research assistants and I made efforts not to pose our questions in relation to Christianity or to be perceived as offering judgments about informants' responses based on Western Christian interpretations. In other words, when people talked of witchcraft or the supernatural we did not "correct" them or allude to historically dominant Western Christian attitudes about these subjects.

Still, we found that when we asked direct questions on spiritually related matters and phenomena typically unfamiliar in professional development settings, such as witchcraft and the supernatural, we almost invariably got evasive answers. One of our principal local contacts, for example, exhibited a level of candor within his home that dissipated when we all sat together with his drinking buddies in the tavern. And a folklorist introduced us to much traditional background in the following manner.

> It is a pity that, you know, we have lost some traditionalism . . . you may not believe [but] these things are there! For example, in Chingola, Kapisha. There is that fountain—spring—that brings out hot water. Our mining engineers—white to be specific—our white men—they wanted to commercialize that kind of water or to domesticate it nicely. They wanted to put some pipes there such that it could be pumped in some homes, used as geezer[18] water. Each time they put that, the pipes could burst. Each—up to now, they have failed. As the Chinese were putting this rail line, TAZARA,[19] there is a place known as Katayi. You know *balya bamulenga wampanga, ba Katayi, kulya kumusumba, kumusumba*, [the spirits like the Mulengas[20] of the Forest, the Katayis, those who reside in royal palaces] meaning some place like but with mysterious powers. They wanted to pass the rail line through that place. And the chiefs warned them but they couldn't believe. They were all bitten by black mambas and they all died. There were eleven of them.[21]

But when pressed by a zealous research assistant as to whether or not the folklorist himself believed, our informant replied that he did not. This same zealous assistant later said that our informant's admission in English was completely counter to the impression he created on a weekly vernacular language folklore radio program.

So, we had much better success asking people in small, casual settings to comment on what they had heard others say rather than what they, themselves, thought. Discussions on spiritual matters and supernatural phenomena were, however, among those we most wanted to have so we made special attempts to audiotape this kind of talk. When we could tape record conversations we asked permission. When tape recording was inappropriate, we sometimes jotted down notes but most frequently taped our own recollections and observations immediately following interviews.

Conversations usually took place in a mixture of Copperbelt Bemba and English. A few interviews were conducted in Chokwe and Lamba. Taped conversations were transcribed and translated by my research assistants, then double-checked by mother tongue speakers of the vernacular languages who were also fluent in English.[22] I have left the translation excerpts that appear throughout this work virtually intact as they were produced even though some points do not conform to strict English grammatical usage. Upon occasion, we were able to return to the speakers and ask for clarification on certain points. This was not done easily, however, since work schedules, transport difficulties, lack of access to technology, and the generally fluid nature of Zambian life did not ensure we would ever see some participants again.

Trying for further clarification of what we heard, toward the end of the fieldwork process we produced a brief summary sheet of salient points. We took these to each of our local contacts and my research assistants explained the concepts in Bemba. Then, we either discussed them on the spot and in the vernacular as needed or left them along with paper, pencils, and self-addressed stamped envelopes for participants to write impressions in letters addressed to me. We encouraged participants to write in Bemba if they wanted to, assuring them that this could be translated. We were aware that written communication of this sort was a less than perfect way to get feedback. But we felt it was worth attempting and did receive a few letters thereafter.

Many Zambian speakers put themselves at considerable risk when talking with us, especially when they offered unfavorable opinions on government and corporate officials. For these reasons, all names in this study except for public figures are pseudonyms.

Information Gathering Processes with Expatriates

Although only one expatriate remained working directly for the mining companies from the heyday of nationalized operations through the privatization era, a number began arriving when the new foreign owners brought in their own managerial staffs. The percentage of expatriates to Zambians, however, was considerably lower than in colonial days, with mine officials estimating that expatriates made up between 1-3% of the overall workforce.[23] Also unlike the colonial days, expatriates in the re-privatized mines were only found in certain managerial and highly technical positions. Day labor or union jobs that in earlier times might have been occupied by either Europeans or Africans remained completely Zambianized[24] in the 2000s. Another change from colonial era privatization was that in the mid-2000s, Zambians sometimes held executive positions and responsibilities. There were also expatriates and Zambians in positions of power and authority at ancillary private companies that supplied the mines with goods and services.

For the most part, expatriates came from Western backgrounds, usually from South Africa or the United Kingdom. No one in a managerial position on the mines proper with whom I talked was an American although a few of the independent consultants and others in managerial positions elsewhere were. Quite a few of the individuals who contributed their insights had spent considerable time underground as engineers or technicians and had then moved into management positions.

The challenge of knowing whom to talk to in the first place was mentioned above, so a word needs to be said about procedures for making contacts among expatriates. Fortunately, one mining official became my chief contact and advisor for this project from the managerial side. With his oversight, I was able to meet a wide range of expatriates in mine management positions. Those initial contacts yielded others through what is sometimes in sociological and anthropological research called the "snowball" technique: one person recommends another person who recommends another to be interviewed. I also already knew a few other expatriates on the Copperbelt; those acquaintances led to other connections in the same way. Through a serendipitous lead arising from archival reviews, I came to know one of the chief independent technical consultants who had spent years in Zambia on World Bank contracts during the mine privatization process. He, in turn, gave me entrée to a number of longer-term Zambian residents (black and white) who provided particularly valuable historical information. In virtually all cases, I was warmly received and many expatriate informants expressed genuine interest in and sympathy for the research project.

Information gathering techniques varied but were generally done under more formal conditions than was the case when talking with Zambians. Interviews usually took place in informants' offices. And, in general, it was easier for me to talk with executives and managers with an audiotape recorder running than it was with members of the general Zambian populace. Informants were always asked permission to be tape-recorded and, in some instances, either refused or asked that the machine be turned off when discussing particularly sensitive material. Such information included candid reflections on historical and cultural issues; details regarding corporate procedures and policies; information on specific interactions with individuals and the like that could put the speakers at professional and personal risk if repeated openly. Usually, I was able to have follow-on meetings with executives and managers to clarify information or ask further questions. A number of managers also provided secondary checks and additional information via email or telephone.

Expatriate sources were promised confidentiality. For this reason, no names or specific titles appear in connection with insights I gained from them.

Historical and Archival Information Gathering

In addition to all the semi-structured interviews conducted, I was able to obtain excellent first hand, or primary source, information in other ways. These included five tours of underground mining operations (four in Zambia's copper mines and one in Canada's coal mines) and three of surface operations as well as attendance at a few key corporate or industry functions.

Considerable historical and technical information for this study also came from important archival sources in Zambia. These included the Chamber of Mines library in Kalulushi; the National Archives in Lusaka; and the ZCCM Archives in Ndola. Many hours spent in these institutions along with monitoring of ZNBC-TV and the local newspapers yielded material that significantly enriched our personal interviews as well as the study.

Study Informant Biographical Sketches

There is one other reason for giving more details about Zambians who participated heavily in this study than about non-Zambian contributors. This concerns the general lack of familiarity with Africa that most Americans have. By offering brief biographical sketches of our principal local contacts, I hope to create a more tangible sense of the real people and surroundings that this investigation sought to understand. The following paragraphs describe the lives of our main Zambian informants as they were at the time we conducted the research and, for this reason, are written largely in the present tense.

Enock Muthwejile, Principal Research Assistant

My principal research assistant was a student in a teacher training program at the Anglican seminary in Kitwe where I taught. His exceptional performance in the classroom as well as his background and demeanor recommended him for work on this project. Enock is Chokwe, in his early thirties, and was raised by his grandparents in a little Copperbelt village. In addition to Chokwe and English, he speaks Copperbelt Bemba and understands Lamba. His grandfather, who is fluent in Portuguese, came from Angola in the 1940s and helped construct many of the mine township houses near Kitwe and Kalulushi.

Enock Muthwejile's village sits about a kilometer off the dust road leading to Chibuluma South mine and is probably 8 kilometers in total from Kitwe. His grandparents, an uncle and aunt, one of his siblings, and numerous cousins still live there. It is a typical Zambian village of mud walled, thatched roof houses. There is no formal land application process so people just settle where they want and move on when necessary.

The family is quite self-sufficient. There is a small well on the premises and a spring about a ten minute walk away. They have several plots of land where they grow sugar cane, maize, potatoes, and other crops. A number of goats and chickens wander around, but it is not certain to an outsider how many of these belong to the family. Not far from the village is a Seventh Day Adventist (SDA) church under construction. Enock and his family are members of the SDAs and his uncle and aunt are very involved in the building project. There is no electricity in the village but Enock always takes his cell phone when he visits so his grandparents can talk with relatives in Lusaka.

Enock tells numerous stories about his childhood that illustrate a traditional upbringing. When he was small, everyone gathered around the fire at night and that was when he learned many customary lessons and sayings. On the way to his first day of school, a snake passed over his foot and Enock's relatives taught him that this was an extremely important sign. During those years, his grandfather would also get up early in the morning to ride his bicycle into town to sell things at market. Enock recalls getting out of bed to help load the bicycle and then push his grandfather up a particularly steep hill at the start of the journey.

After attending secondary school, Enock went to Ndola where he has some relatives and then to Chingola where he operated various small businesses and even a preschool. While in Chingola, he ran for a city council position and almost won. Following completion of his teacher's training, Enock started an after school tutoring program in Ndola before working with this project.

Lubasi Mwangala, Research Assistant

Lubasi was the youngest member of the teaching staff at the seminary where I taught. While a student at that same institution, he had been identified for his outstanding academic abilities and so moved into the staff ranks following graduation. Lubasi is in his late twenties and grew up primarily in Livingstone. His father is Lozi and his mother is Tonga so Lubasi speaks both languages along with Copperbelt Bemba and English. Because of the family's mixed heritage, they commonly spoke English in their home. His parents are teachers and, when Lubasi was growing up, his father served as a headmaster in Livingstone and Serenje where the family lived for a few years. They are also quite prominent leaders in local Anglican Church activities.

This urban, academic, comparatively comfortable background inspired Lubasi's own pursuits. It gave him few formal teachings in traditional African beliefs, but he learned quite a bit from childhood playmates. He also learned early about economic and social stratification when he went to school with peers who were significantly less well off financially than he. After secondary school, Lubasi entered seminary to study for the Anglican priesthood. Shortly after ordination he also completed a bachelor's degree in theology.

Roger Chaba, Mufulira

My research assistants met Roger through a mutual friend, Wilson. Wilson owns a township shop where we spent time drinking soft drinks and people watching. Some interactions with Roger also took place outdoors near a large tree we nicknamed our *insaka*[25] tree and in local pubs.

Roger has what is sometimes referred to as a Person-in-Charge, or PIC job, underground. As a casual employee of a mining company, he oversees arrangements for various supplies that are hauled up and down and throughout the mine. Though only in his twenties, he has been working on the mines for a few years and occasionally lets his friends know when a mining job comes open. But many of them would rather stay on surface even without the better salary that underground jobs offer. Roger himself wants only to work underground for another year or two and then go into the chicken rearing business.

Roger lives in a better section of one Mufulira township. In Mufulira, more so than the other towns, the distinction between compounds that were owned and run by government and those that were owned and run by the mines is obvious. Houses in the mining compounds were somewhat larger than government houses and better maintained. Even though everyone has difficulty keeping things up these days, the historical difference between these two sections is still apparent. One thing everyone near the mine site shares, though, is the effect of the old smelter. Areas closest to the mine are arid; what grass, trees, and flowers do survive are stubby and straggly looking. Sometimes, depending upon wind direction, it becomes difficult to breathe.

Frederica Chipongoma, Luanshya

Frederica is in her early forties and a counselor for one of the local high schools where she has responsibility for several hundred students. She also teaches English composition. Her husband is a mechanical engineer by training and worked in the mining industry for 15 years before being retrenched under privatization. Now he works at a training institution in Ndola that contracts services for the mines. Since daily commuting from Luanshya is very expensive, he frequently stays in Ndola during the week, returning home on the weekends.

The Chipongomas live in what used to be a European section of town. The streets are wide and paved, even if pot-holed, and the brick houses are considerably larger than township homes. Each house has a substantial yard and many also have what were servants' quarters out back. Frederica was born in Luanshya into a mining family and grew up near where she now lives. By the time she came along, that section of Luanshya was already racially mixed with African families being scattered about among European and Indian families. Frederica can remember when machines owned by the mines kept the streets swept and water pumped from the underground made lawns green. Back then, the yards would have been well tended and filled with flowers. The hardiest plants from that era still survive in many places and, with only a little imagination it is possible to see in the area a modest middle class neighborhood similar to those in Britain or the United States.

Frederica is extremely involved with her church. Originally Roman Catholic, she joined her husband in the Anglican Church when they married. Today, her leadership abilities mean she is put in charge of many things and a local priest who knew of her responsible nature recommended her as a contact for this study. When we went to Luanshya, Frederica was always well prepared with a slate of interviewees for us to speak with or parts of town for us to tour.

Matthew and Martha Kapumba, Kitwe

The Kapumbas live in a township near Kitwe. Matthew is in his early fifties, from Northwestern Province, and spent his youth in Mufulira. Martha's family comes from the Lundazi area in Eastern Zambia, but she was born on the Copperbelt and went to school near Kitwe. The Kapumbas have a number of children, the oldest of whom is a miner. Matthew's father lives with them as well.

Matthew began work on the mines as a lasher in 1975. He describes this job as "very tough," the worst of all. When blasting takes place, the work progresses at about two meters per blast. Lashers go in and physically remove—with shovels, picks, and wheelbarrows—the rocks and rubble created by the blast. After working as a lasher for about a year, he had other short-term assignments and then spent 18 years as a driller. Today, Matthew is a shift boss for one of the mines and has approximately 600 men for whom he is accountable.

The Kapumbas have a small home that they bought when President Chiluba sold miners' houses to them. Now they are adding on to the house as they can since Matthew will retire there in a few years. It is a typical township dwelling that sits within a grid-like pattern of dust lanes a few kilometers from the town center. These homes are built of concrete blocks with tin roofs. Each one is fenced in by hedgerows that have grown so high it is really only possible to see the tops of the dwellings from the lanes. Inside the hedgerow, there is a neat yard. A water tap runs next to the house and empties into a concrete basin on the ground. This is where the family does dishes and laundry. The front door leads into a kitchen that has an electric stove but no refrigerator. Blue paint peels off the walls of the living room and bits of sheets hang in the windows acting as curtains. At night, one bulb suspended from the ceiling provides light. A color television set, frequently switched on, sits in one corner. There are a few pubs nearby where Matthew goes from time to time. Upon occasion Martha will join him in drinking beer. Matthew does not, however, seem to spend a great deal of time at these places since when he is not at work, he is often doing things for his church or visiting in the homes of friends.

Matthew and Martha are both leaders in their local Anglican parish. Martha is a prominent Mother's Union member who coordinates activities such as cooking meals for women in the local prison and hosting members of other churches for choir competitions. Matthew is a member of Men's Anglican Fellowship. We met the Kapumbas through the church when Enock made an announcement asking for volunteer study participants. Our subsequent interactions with them have been many and pleasant, occurring at church, over meals in their home or mine, and a few times in local drinking establishments.

Sunday and BanaAlice Mutale, Chingola

The Mutales are close friends of Enock's. They live in one of the townships and Sunday works as a section boss for the local mine. He is in his late thirties. On the Copperbelt it is often customary to call parents not by their given names but by the name of their oldest child preceded by either Bashi for a man or Bana for a woman. So Sunday's wife goes by the name BanaAlice, after their first born, a young girl who has just completed secondary school.[26]

The township in which the Mutales live is similar to that of the Kapumbas but in somewhat better shape. The roads are dust but wider than in the Kapumba's compound and the houses are larger. Most also have small yards with grass and flowers in front, vegetable gardens out back. Since miners now own these homes, their conditions vary according to the occupants' abilities to keep them up. So it is quite common to see very nice, tidy dwellings next to those that are run down.

This is the case in the Mutale's neighborhood where their house is one of the best kept around. The living room has a couple of large windows that let in lots of light and the walls are painted bright yellow so the whole effect is open and cheery. Furniture jammed into the small space consists of red plush sofas and two chairs, all accented with embroidered doilies. In the middle is a coffee table and a large china cabinet is shoved against one wall. This holds the television set, video recorder, and working telephone with answering machine, various dishes and supplies. The Mutale's kitchen has a stove and sink but no refrigerator. Two bedrooms and a bathroom are located behind the kitchen.

Sunday's family comes from Eastern Province, but his father moved to the Copperbelt for a mining job and stayed with his sister who was married to a miner. BanaAlice's family was from Northern Province but her father, too, had come to work on the mines and so both Sunday and BanaAlice were born in Chingola. BanaAlice's father worked as a driver in the huge open pit mine and occasionally she would see him in his truck as she walked past to school. Once, her grandmother also saw him driving when the truck was very full. She kept saying, "That's too much for him to carry," thinking that her son was using his own energy to haul all that material. The Mutales are members of the United Church of Zambia (UCZ) and BanaAlice sings in the choir. They have five children, the youngest of whom is a toddler.

Sunday chose to involve himself in this study by regularly organizing *insakas* with many friends and co-workers. He routinely gathered anywhere from 8 to 20 neighbors (a few of whom became regular attendees) for animated discussions which we tape-recorded. These gatherings were populated completely by men until, for the penultimate meeting, they suggested that women be invited.

Earlier and Additional Copperbelt Studies

This study is not the first to look at work relationships in Zambia or the Zambian mining community, but it appears to be the most thorough cultural analysis of the situation following full re-privatization of the mines in 2000. Further, while a number of previous Copperbelt studies have touched on matters of religion, meaning, and reality, they have principally done so using anthropological and sociological analyses without engaging matters of religion as this study does. So, this study is attempting a somewhat different perspective to suit the times but is also, in a sense, carrying forward bits and pieces of hard work and insights contributed by those who have gone before.

Of numerous works that have chronicled various Copperbelt matters over the past decades, the most influential on the present effort are the following. Raymond Buell's 1928 work contains valuable historical information on the early years of the British South Africa Company (BSAC).[27] In the early 1930s, J. Merle Davis produced *Modern Industry and the African*[28] as a result of concerns about the overall impact that rapid industrialization was having on Africans. His was also an attempt to analyze the situation from an interdisciplinary standpoint, specifically sociology, economics, politics, and religion.[29] Walter Cline's late 1930s doctoral dissertation on metals and metal working techniques in the Copperbelt-Katanga (DRC) area records significant information about traditional mining and smelting practices before industrial mining had so taken hold.[30] Andrew Roberts's historical work was particularly valuable for informing this study's review of Zambia's early industrial era developments.[31]

The colonial and early independence eras produced quite rich material. In the 1950s, Hortense Powdermaker's anthropological investigation recorded in her book, *Copper Town: Changing Africa*, documented much about African life as the shift toward independence was progressing throughout the continent.[32] Around the same time, William Watson examined the effects of instituting cash economy values among communities of Africans who came to work on the mines.[33] A. L. Epstein considered politics and urbanization trends on the Copperbelt for a number of years from the 1950s onward, producing two books about the province and its people.[34] These works helped inform the studies of several others: John Taylor and Dorothea Lehman produced *Christians of the Copperbelt* based on research conducted in 1958. Richard Hall's 1965 book simply entitled *Zambia*[35] was one of the first comprehensive analyses published after political independence and it, in a vein similar to Davis, took a broadly interdisciplinary stance. Also during the 1960s, President Kaunda's speeches, letters, and essays described a clear philosophical vision for governance of the now politically independent nation. The mines inevitably played a prominent role in how the government went about its business and the record of his thinking on all this has lent a crucial dimension to the study.[36] President Kaunda also wrote the foreword for a related book, Henry Meebelo's *Reaction to Colonialism*[37] that, while focusing on the Northern rather than Copperbelt Province, provides important information about political and religious movements that were significant for the independence struggle. The business of the mines received particular attention from other quarters in the 1970s. Michael Burawoy produced *The Colour of Class on the Copper Mines: From African Advancement to Zambianization*, a study that examines the controversial process of moving Zambians into mining jobs previously occupied by European expatriates.[38] That same year, Antony Martin wrote *Minding Their Own Business: Zambia's Struggle Against Western Control* describing a range of constraints the Kaunda government faced early in its administration.[39] A work that reflects shifts toward American-led international development policy making and planning is Irving Kaplan et. al.'s *Area Handbook for Zambia.*[40] Its being published by the United States government suggests an intended audience of foreign service personnel and it is further noteworthy for attempting an interdisciplinary perspective.

The number of comparatively recent works is more limited than those produced during the time that a European presence on the Copperbelt was more obvious. Although published in the 1990s, James Ferguson's *Expectations of Modernity: Myths and Meanings of Urban Life on the Zambian Copperbelt*[41] is based upon fieldwork he conducted in the late 1980s. Ferguson documented a time of decline in the mines' history just a few years before President Kaunda was voted out of office and the idea of re-privatizing state owned enterprises gained increasing prominence.

From Ferguson's examination until now, the most detailed studies of the mines appear to be predominantly economically based or grounded in historical, labor, and political studies. Miles Larmar's *Mineworkers in Zambia: Labour and Political Change in Post-Colonial Africa*[42] covers some of the same geographic area and time period as does the present work but it takes primarily an historical approach to labor relations. John Craig's journal article, "Putting Privatisation into Practice: The Case of Zambia Consolidated Copper Mines Limited" is a sensitive analysis primarily of economic constraints under which the Chiluba government worked.[43] Reports such as "For Whom the Wind Falls: Winners and Losers in the Privatization of Zambia's Copper Mines"[44] and "Undermining Development? Copper Mining in Zambia"[45] do involve some religious concerns both through sponsorship by faith-based organizations and by providing important insights into the overall situation. The World Bank and USAID also continue to issue documents and reports on Zambia.

Study Parameters and Limitations

It is necessary to make just a few more points about what readers can expect to encounter before proceeding with the Copperbelt narrative. As suggested above, because this work is posed from a theoretical standpoint that suggests "reality" can differ radically depending upon context, readers should not expect a one-to-one correspondence between the ways in which Zambians thought about key issues and the ways a dominant development narrative posed them. In fact, not even the issues themselves should be expected to correlate.

Additionally, substantive differences between the manner in which information was typically gathered and disseminated within Copperbelt communities and from what might reasonably be expected in Western settings deserves mention. Using stories and anecdotes (rather than surveys, experiments, and other sorts of quantitative research methods) did not make the Zambian information any less credible; it simply documented a widely acceptable way of communicating that differed from prevailing norms in America.[46]

Finally, while sympathetic effort has gone toward vantage points of ordinary Zambians, in no way should this study be construed as a wholesale indictment of American or Western culture. Nor should it be viewed as an unqualified commendation of alternative views posed by the understanding of Zambian Copperbelt residents. It is principally an argument on behalf of thoughtfulness.

Notes

1. See especially the discussions of how Powdermaker and Taylor and Lehmann conducted their own Copperbelt investigations. Hortense Powdermaker, *Copper Town: Changing Africa, The Human Situation on the Rhodesian Copperbelt* (New York: Harper and Row, 1962). John V. Taylor and Dorothea Lehmann, *Christians of the Copperbelt: The Growth of the Church in Northern Rhodesia* (London: SCM Press, 1961).

2. Largely because of the dominance that economic theory has exercised over the shaping of development thought, there has also been an overwhelming reliance on quantitative data research and collection as opposed to qualitative research and data collection. Historically cross-cultural qualitative work has been considered the purview of anthropologists, sociologists, and religious workers. It is for this reason that the preponderance of linguistic and cultural analyses of non-Western regions produced from the seventeenth century until roughly the 1960s was the work of Christian missionaries. In a study of the Copperbelt that was conducted soon after full scale mine industrialization had begun, Merle Davis recommended that missionaries in particular be trained in an interdisciplinary manner to include instruction in anthropology, sociology, economics, community planning, medicine, and so on. Davis, *Modern Industry and the African*, 391. See also Robert, *Christian Mission* on these matters generally.

3. The issue of trustworthiness is crucial to qualitative research studies and candor is not as easily obtained as may be wished or expected. Some of the problems encountered in the Copperbelt setting had deep historical roots that researchers have commented upon for decades. More about this matter is contained in chapter seven.

4. Robert Chambers contends that preoccupation with quantification and measurement has so affected the development industry that the existence of data is frequently more important than its accuracy. *Whose Reality Counts? Putting the Last First* (London: Intermediate Technology Publications, 1997), 40.

5. The methodology employed for this study has similarities to other Copperbelt studies such as Taylor and Lehmann and Ferguson.

6. While what constitutes a working class wage or standard of living can be calculated and construed in a variety of ways a standard, union mining job was yielding take home pay of approximately $300 per month or $3,600 per year in 2007. Abi Dymond, Kato Lambrechts, and Simon Chase, "Undermining Development? Copper Mining in Zambia," Report sponsored by Action for Southern Africa, Christian Aid, and SCAIF (2007), http://www.actsa.org/Pictures/UpImages/pdf/Undermining%20development%20 Development%20report.pdf 2007 (accessed March 23, 2009) contains a case analysis of wage structures and cost of living for miners working at Konkola Copper Mines. By contrast, salaries for expatriates in the $180,000 to $200,000 per year range were commonly being advertised at the time of this writing. See, for example, information contained on minejob.com, a website that posts international openings in mining and geology, http://www.minejob.com/keywords-Metallurgy.html (accessed December 5, 2009). On

the Copperbelt proper, union reports of salary disparities recorded approximately $500 per month for local miners as compared with $6,000 per month for expatriate Indian employees of Konkola Copper Mines in late 2009. Chiwoyu Sinyangwe and Ernest Chanda, "Appreciate Workers' Efforts, Nchanga MP Urges KCM," *The Post*, November 16, 2009, http://www.postzambia.com/post-read_article.php?articleId=1945 (accessed December 5, 2009). These numbers accord with figures that were informally communicated by study participants during the years of field work for this project (2003-2005).

7. Exact ownership of mine financing companies is not always easy to discern. Roan Antelope, renamed Luanshya Copper Mines, was majority owned by Enya Holdings, BV, which, some sources listed as a British company, despite its registration in the Netherlands. "Enya Holdings to Operate Zambia's Top Cobalt Producer," Miningreview.com, June 26, 2009, http://beta.miningreview.com/node/15730 (accessed December 5, 2009).

8. "Survey of the Zambian Mining Industry 1995-2004" (Chamber of Mines of Zambia, Kalulushi, October 2005), 7.

9. Yulai Chen, "China in Africa" (presentation delivered at African Studies Center, Boston University, Boston, MA, November 23, 2009). See also Moyo, *Dead Aid*, 98ff.

10. There was a particular history of suspicion on the Copperbelt and in Zambia that informed our procedures. Taylor and Lehmann report that their initial entrée into the African mining communities was made possible by mine company and government officials, all of whom at the time would have been European. They state that this was the only way they could gain access to the people they wished to meet. But as soon as they could, they worked to minimize their "connection with authority in any form." Taylor and Lehmann, *Christians of the Copperbelt*, viii.

11. This is not a new phenomenon for the Copperbelt. In her historical reconstruction of experiences relative to both the Northern Rhodesian and Belgian Congo copper mining regions, White observes that the complexities of those workplaces "might best be reconstructed from sources officials did not take seriously, stories considered the domain of 'ignorant natives,'" *Speaking with Vampires*, 280.

12. This can be understood as a personal manifestation of the sort of power differentials that Wuthnow observes in relationships between rich and poor countries. Wuthnow, *Boundless Faith*, 58.

13. Paulo Freire, *Pedagogy of the Oppressed*, 30th anniversary ed. (New York: Continuum, 2000).

14. Recognition that this sort of viewpoint disparity can have a major impact on research findings appears to be behind the World Bank study, "Voices of the Poor," released in 2000. Deneulin notes that this study was a "landmark in the literature on poverty" because, rather than relying on statistical surveys designed by economists, the research effort "asked poor people themselves how they experienced poverty." Deneulin goes on to remark that one of the study's "most interesting findings is that religion permeates people's conception of well-being." Deneulin, *Religion in Development*, 43. See also the study, http://go.worldbank.org/IM4HTOJ690 (accessed December 5, 2009).

15. Such an "allergy" reached back at least to the time in which Taylor and Lehmann were working on the Copperbelt. They report encountering problems when they tried to circulate a survey in the Nchanga area that was designed to elicit statistical information on church membership and practice. But the survey prompted rumors that the motive behind it was really to identify Africans who were supportive of Federation (covered in chapter three.). This prompted people not to participate and the data that Taylor and Lehmann finally collected in this regard they deemed useless. Taylor and Lehmann, *Christians of the Copperbelt*, viii.

16. It is possible that this uniformity of viewpoint was a manifestation of characteristics identified by Taylor and Lehmann. In a discussion of the development of African self-expression, they suggest that African miners' increasing ability to give themselves voice despite existing in a situation of radical power disparity came about as an unintended consequence of the industrial setting itself. While noting that "tribalism" as a force was still evident at the time of their field work (1958), they found that it had waning authority in certain social settings. "A man did not forget that he was Bemba or Nyanja by tribe, but the fact was not germane when he needed to express himself communally with other employees to the management, with other town dwellers, to the local authorities, or with other Africans to the white rulers." Taylor and Lehmann, *Christians of the Copperbelt*, 130-31.

17. Copperbelt Bemba is a mixture of Bemba, other Zambian languages and English, reflecting the cultural mix of the region. It is notably different from "pure" Bemba.

18. I.e., hot water heater.

19. Tanzania-Zambia Railway Authority (TAZARA) refers to a rail line built between Kapiri Mposhi in Zambia to Dar-es-Salaam in Tanzania. The line was constructed in the 1970s by the Chinese as a way for land locked Zambia to circumvent import-export difficulties posed by the liberation war taking place in Northern Rhodesia/Zimbabwe.

20. The translator notes this refers to the Bemba deity.

21. Copperbelt folklorist and radio personality, interview with author, 2004.

22. Exceptions are the Lamba translations. These were checked by a mother tongue Bemba speaker who also speaks Lamba. The two languages are closely enough related that my research assistants could ask questions in Copperbelt Bemba and informants could answer in Lamba.

23. A human resources departmental manager at one of the mines reported an average of 1% expatriate employment from the point of that company's privatization through early 2006 and estimated that all together the expatriate workforce on the Copperbelt during the early- and mid-2000s would have been no higher than about 3%. In the mid-2000s, expatriates came not only from Commonwealth countries, but also from Latin America and Asia. Very few were American.

24. This term and the process it designates are discussed more fully in the next chapter.

25. In traditional Zambian societies, *insakas* were meeting places where village men would gather to discuss issues of community life. In mid-2000s Zambia, the term also often applied to small, thatched huts or open shelters at restaurants and beer halls where people could drink and talk.

26. If married couples are childless the designation changes. Prior to Alice's birth Sunday and his wife would have gone by the names BaMutale and BanaMutale.

27. Raymond Leslie Buell, *The Native Problem in Africa* (Cambridge, MA: Bureau of International Research of Harvard University and Radcliffe College, 1928).

28. Davis, *Modern Industry and the African.*

29. Davis, *Modern Industry and the African*, xxvi.

30. Walter Cline, "The Sources of Metals and Techniques of Metal Working in Negro Africa" (PhD diss., Harvard University, 1936).

31. Andrew Roberts, *A History of Zambia* (New York: Africana Publishing Co., 1976).

32. Powdermaker, *Copper Town.*

33. William Watson, *Tribal Cohesion in a Money Economy: A Study of the Mambwe People of Northern Rhodesia* (Manchester: Manchester University Press, 1958).

34. A. L. Epstein, *Politics in an Urban African Community* (Manchester: Manchester University Press, 1958) and *Scenes from African Urban Life: Collected Copperbelt Papers* (Edinburgh: University of Edinburgh Press, 1992).

35. Hall, *Zambia.*

36. See, for example, Kenneth D. Kaunda, *A Humanist in Africa: Letters to Colin M. Morris* (London: Longmans, Green, and Co., Ltd., 1966).

37. Henry S. Meebelo, *Reaction to Colonialism: A Prelude to the Politics of Independence in Northern Zambia 1893-1939* (Manchester: Manchester University Press, 1971).

38. Michael Burawoy, *The Colour of Class on the Copper Mines: From African Advancement to Zambianization*, Zambian Papers No. 7 (Lusaka: University of Zambia Institute for African Studies, 1972).

39. Antony Martin, *Minding Their Own Business: Zambia's Struggle Against Western Control* (London: Hutchinson and Co., 1972).

40. Irving Kaplan et. al., *Area Handbook for Zambia*, 2nd ed. (Washington, DC: U.S. Government Printing Office, 1974).

41. Ferguson, *Expectations of Modernity.*

42. Miles Larmar, *Mineworkers in Zambia: Labour and Political Change in Post-Colonial Africa* (London: I. B. Tauris and Co., 2006).

43. John Craig, "Putting Privatisation into Practice: The Case of Zambia Consolidated Copper Mines Limited," *The Journal of Modern African Studies* 39, no. 3 (September, 2001): 389-410.

44. Alastair Fraser and John Lungu, "For Whom the Wind Falls: Winners and Losers in the Privatisation of Zambia's Mines," Report sponsored by Civil Society Trade Network of Zambia and Catholic Centre for Justice, Development and Peace, Zambia Episcopal Conference (2006), http://www.minewatchzambia.com/reports/report.pdf (accessed March 23, 2009).

45. Dymond, Lambrechts, and Chase, "Undermining Development?"

46. Stephen Ellis and Gerrie Ter Haar, *Worlds of Power*, 7. See also Gilkey, *Nature, Reality, and the Sacred*, 80 and Rasmussen, *Earth Community Earth Ethics*, 32.

♀

Chapter Three

Public Stories of Zambia's Mining History

A proverb common to Southern Africa says, "You can't know where you're going unless you know where you've come from." This proverb well acknowledges the human need for fashioning coherence within our daily lives.[1] It makes overt the connection between context and action. Indeed, how we meet the need for coherence has an impact on whether understanding is achieved or thwarted, since what might be considered relevant within one realm may not be thought of similarly in another.

Likewise, the trajectory of Zambia's development process can be charted differently depending upon who is telling the story. According to a more public narrative, the starting point for privatization of the mines and other nationalized companies concerned the concept of resources and how they were used.[2]

Pre-colonial and Colonial Era

Archeologists and linguists have enough evidence to suspect that Bantu speaking peoples have been living in what we now call Zambia since the Stone Age.[3] Zambia today has over seventy groups or tribes with those traditionally thought to have settled on the Copperbelt being Lamba.[4] While this information might seem merely an historical fine point it proved quite useful when Zambia became a nation-state in 1964. That part of the story comes later.

Copper runs like an arc from an easterly region around Ndola up into what is now Katanga Province, DRC, back down through the Copperbelt and on out toward Kansanshi and Lumwana in the Western Province. Evidence of major mining activity reaches back to the first millennium BCE and copper workings found out west at Kansanshi and east at Bwana Mkubwa near Ndola are among the oldest on the continent. Estimated dates for these and similar finds vary from the fourth to eighth centuries BCE.[5] What is more, the early mining efforts of Africans in this region were intentional and sustained. Near Bwana Mkubwa, for example, miners worked a vein of high-grade copper ore down to one hundred feet and as far as twenty to thirty feet across.[6]

Much of what Africans mined they also exported. Their earliest trading partners were probably Arabs and by the eleventh to fifteenth centuries their trade routes were well developed.[7] These routes linked Africans with the Portuguese who, by the 1400s, controlled both southern coasts of the continent and had established a busy sea route to Goa, India.[8] Eventually, the copper trade became intertwined with the missionary movement and slave trade. Arab slavers dealt in copper exchange; slave convoys carried out copper bound for export.[9] Copper alloy made into brass products was re-imported for sale, sometimes returning to Africa on the same ships that brought European missionaries.[10]

Many of these Christian missionaries came to Africa infused with a confidence and curiosity similar to that experienced by their contemporary European commercial traders who visited the continent. This confidence had some practical as well as theoretical origins. Industrial era inventions of electrical power and steam engines had given the means for Europeans to embark on previously unimaginable journeys.[11] Their serious religious environment gave them impetus to engage in far reaching efforts that included evangelism but also were often infused with genuine desires for service.[12]

For example, at that time missionary work in Africa took on a special urgency due to the slave trade. This was particularly true for one remarkable person who also happened to encounter Central Africa's mining culture. During the early to mid-1800s, Scottish missionary, explorer, and physician David Livingstone became famous in England for his adventures. Livingstone believed that a water trade route across the continent would open the way for Africans to set up workable alternatives to the slave trade. While he apparently never converted a single African to Western Christianity, Livingstone's passionate belief that the best means of ending the slave trade was through introducing commerce and Christianity inspired scores of Western missionaries to go to Africa.[13]

His association with the Zambian region was especially multi-dimensional. Livingstone's books and journals reveal an extensive familiarity with copper mining activities in what is today the nearby Katanga Province. And when suffering from foot ulcerations he recorded that Africans advised him to use a combination of malachite, water, and stone applied with a feather. It seems to have been a painful, but effective, remedy.[14] Livingstone ultimately died just somewhat east of the Copperbelt without ever having found a transcontinental water route. But his work and inspiration did—long afterwards—help stop organized trade in slaves.

Another famous European whose legacy will forever be associated with Zambia also came to the Copperbelt by way of Katanga. English entrepreneur Cecil Rhodes sent prospectors up from his adoptive home of South Africa in the late 1800s with orders to search for a gold deposit comparable to the huge Witwatersrand. Prospectors never found significant gold reserves north of the Zambezi River, but through a series of highly questionable and scarcely honored treaties with local African rulers, Rhodes's British South Africa Company (BSAC) eventually took over a large region to use as a buffer zone against Belgian expansion from the north.[15] This territory represented the northernmost English outpost connected to South Africa. Rhodes also considered the area's Africans as a superb labor force for the Belgians' Katanga mines and his own in Southern Rhodesia.[16] For over two decades beginning in 1899, the BSAC administered the region first in two sections (North Western and North Eastern Rhodesia) and then one amalgamated Northern Rhodesia before turning responsibility over to the British Government in 1924.[17]

By the time Rhodes and other Europeans had begun prospecting in the region, traditional African mining and smelting practices had almost completely died out. Decades of slave trading had decimated local populations and the disruption caused by that trade left any remaining communities highly unstable. Additionally, the chemically complex Copperbelt ore bodies were mostly buried underground, making them harder to find. Out west at Kansanshi and back east at Bwana Mkubwa there were at least old workings and malachite green stained hills sticking out of the landscape to see. But industrial era outsiders looking for metal workings or mineral deposits on the Copperbelt proper would not have detected much. Sometimes they used scientific means to find copper and sometimes they simply resorted to wandering around with pieces of malachite in their hands asking locals if such rocks looked familiar.[18]

At least two points about this history are worth keeping in mind in relation to contemporary development policy thinking. The first point concerns the *nature* of Western entities involved with Africa; the other concerns *expectations*. To the first point, modern day development initiatives are often publicly discussed in the context of nation-states with development agreements being drawn up as if among equally sovereign countries. But the BSAC was a private company managing a place Westerners called "Northern Rhodesia" and, as such, invested comparatively lightly in the sorts of infrastructure projects typically assumed as helping to constitute independent nation-states. These include roads and bridges; water and power delivery systems; social service institutions; standing armies; treasuries; and the like. In fact, it was generally the Christian missionaries who took care of such things as schools and hospitals since company officials did not see these as their responsibility. Further, the BSAC's objectives were as profit oriented then as is any comparable private enterprise now. Records from the early 1900s, for example, describe activities in the region in economic and managerial language that would sound familiar in virtually any twenty-first century corporate board room. The Company, for example, was known to chafe at government regulations and uphold shareholder satisfaction in arguably much the same way that private corporations do now.[19]

Second, it is crucially important to consider expectations when outside interests are involved with the continent. Modern day development practices often favor initiatives that promote rapid self-sufficiency and sustainability. A philosophical viewpoint underlying such policy seems to be the assumption that self-sufficiency and sustainability are natural habits of well run nation-states. In other words, self-sufficiency and sustainability both should be an objective achieved fairly quickly after certain sets of actions are taken. But Zambia's history indicates that the road to sustainability in the colonial era was understood as long and the relationship between private companies and public governance somewhat different from what exists today. For example, while Great Britain did have a presence in Northern Rhodesia before officially taking it over as a protectorate, in those early days, the government relied upon subsidies from the BSAC to fund its operations. Further, during the entire period of BSAC administration expenses needed to run operations on the ground always exceeded income; Northern Rhodesia was not self-supporting for the quarter century it existed as a Company administered place nor was it for some time thereafter.[20] While Company and government officials might have desired Northern Rhodesia to pay its own bills, in actuality it did not and could not for years.[21] More than a decade into the BSAC arrangement, investors were still searching for the proper strategic plan that would maximize economic profits even as disillusionment with the territory's potential had already set in.

Meanwhile, the efforts of those existing on the ground in Northern Rhodesia were often keenly focused on survival of a different sort.[22] Prospectors and adventurers with a taste for the chaotic, dangerous, and occasionally romantic, explored the bush setting up rough outposts near promising sites. Certainly, some went to the region motivated solely by the possibility of finding great material wealth. But many other Europeans were attracted to the locale for the manner in which it differed from the rainy British Isles. Spectacular sunrises, sunsets, and months of clear, dry weather could be almost intoxicating to expatriates. Yet they did not outweigh the hardships of sheer living in those early days. While Africans had contended for countless generations with mud, malaria, and malnutrition, unaccustomed Europeans often fared quite poorly.[23]

Those who did prevail eventually compiled enough exploratory evidence to suggest that the Copperbelt's potential was enormous. Cheered by this news, financiers organized two major companies that would eventually control Zambia's mining industry from the late 1920s until the 1960s. One company, Rhodesian Selection Trust (RST), was funded largely out of the United States and United Kingdom. The other, Rhodesian Anglo American Corporation, was a subsidiary of Ernst Oppenheimer's Anglo American Corporation (AAC). It also had American and British funding but was based in South Africa.[24] Despite the relative cultural uniformity of their backers, these two companies were very differently run with RST taking a more relaxed, open managerial style and AAC assuming hierarchical, formal ways of doing business. Zambians with long-term experience on the mines could easily point out remnants of these varying corporate cultures well into the mid-2000s.

Constructed along industrial lines (rather than according to their precolonial configurations), most of the major Copperbelt mines, along with Kansanshi in Western Province, opened for commercial production during the late 1920s and early 1930s. Some fairly promptly closed again when the Great Depression dampened the industrialized world's demand for copper.[25] But the mines reopened as conditions improved and openings and closings formed a familiar pattern with the economic vacillations of the next several decades.[26]

From the start of the mines' intensely industrial period, Africans came in droves out of the rural areas to work. One way of looking at such a major migration from the countryside into town would be to see it as evidence that Zambians, every bit as susceptible to the attractions of wealth and material goods as anyone else, gladly accepted the opportunity of engaging in salaried work. Such an analysis might assume that Zambians all came to the mines because they wanted to. But the situation was actually more complex than that. Africans had to seek work on the mines because colonial government policies necessitated it. Pressure for wage work was particularly created by the hut tax, a plan the British government regularly instituted in its colonies to support local administrative activities.[27] Under this system Africans were assessed a tax that had to be paid in cash. But since the rural areas operated outside the principles of cash economies, men had to go to work in places where they could be paid in monetary form. Proponents of such taxation praised it as a way of involving the local people in paying for services and amenities located within their areas such as roadways, schools, and hospitals.[28] Yet they apparently overlooked the tax's impact on local cultural values. Dealing in cash instituted a new authority for Africans that undermined their traditional chiefs' authority and disrupted previously established economic systems.[29] Moreover, male workers' migration to urban areas interfered with long standing rural life cycles. Farmers were forced into absences at key times of the agricultural year when they engaged in stints as miners underground.[30]

In light of destabilization created by shift schedules, wage work, and imposition of cash economy taxes, it is also worth noting the ways in which Africans countered to achieve equilibrium. With remarkable speed, they learned to organize their communities and take action where they could. Within the limited opportunities offered them, they chose whom they worked for and founded support structures outside and inside the workplace.[31] One well-documented strategy was, according to outside observers, a spontaneous movement among African miners to form Christian associations across denominational lines. There were no European missionaries assigned to the Copperbelt's indigenous populations at the time. But by 1925, an autonomous African congregation in the compounds of Ndola had achieved sufficient strength to elect a board of elders and nominate candidates for baptism. For more than half a dozen years, this movement flourished completely under African leadership yielding a successful school, outreach activities, and construction of a church building.[32] Not far away, near Kitwe, miners were also meeting ecumenically for the purpose of worship and skills building that would help them adapt to town life. The United Missions in the Copperbelt (UMCB) was a collaboration between various European mission agencies, the mining companies, and indigenous congregational leaders. Schools, sports, and support structures for women were all part of this effort that was ultimately institutionalized in the Mindolo Ecumenical Centre.[33]

Within a decade, African miners had become so well informed and coordinated that they were able to create a notable series of what official reports labeled "disturbances" in Luanshya and Kitwe regarding wage and labor inequities.[34] Historians and anthropologists, however, in extensive analyses of the 1935 riots have termed them the first important demonstration of industrial unrest in Northern Rhodesia[35] even though evidence exists as far back as the 1910s that Africans attempted organized resistance to economic exploitation.[36] In 1935, the British government responded by expanding its tribally represented judicial and legislative systems and also instituted a labor department. Colonial officials faulted, at least in part, a secretive, emotionally charged and highly politicized movement called Watch Tower as a predisposing factor in the events. This was due, they said, to the movement's contempt for civil and spiritual authority especially "native authority."[37] Contemporary scholarly consensus sees this as misreading the movement but such blame placing illustrated official unease with the entire context in which Europeans and Africans existed together.[38]

By the late 1940s, Africans at all the major mines had formed official labor unions but the workforce's increasing sophistication and development of the mines themselves were not matched by regional improvements.[39] In fact, Northern Rhodesia's infrastructure expansion was consistently under funded as the British government and the BSAC extracted revenue from the Copperbelt while returning only a fraction for local use.[40] Had World War II not intervened, this might have been somewhat corrected, for such was the recommendation of consultants brought in to assess the territory's running.[41] As it happened, though, debate over development initiatives was pre-empted when war made the increase of industrial copper production a more urgent policy matter than social service provision for those directly involved in mining and processing.[42]

The war years and immediately thereafter eventually proved to be the time of greatest expansion for Zambia's copper industry. Northern Rhodesia's European population rose to its highest levels while employment on the mines grew substantially and quickly.[43] These influxes and an economically optimistic production environment prompted a return to thinking about overarching development issues. Major infrastructure for both the European and African sections of Copperbelt towns was constructed during this time. The resultant towns and cities were of sizes and complexities unimaginable across most of the continent, facilitating a transition to what eventually would become the most urbanized country in sub-Saharan Africa.[44] For expatriates, the auxiliaries of mine life were staggering, representing the best of Britain in the bush. Even mid-level European workers could expect to enjoy scrupulously maintained sports facilities, up to date cinemas, and schools and hospitals of the highest quality.[45]

For Africans, the situation was somewhat different. The alternating rural-urban work patterns they had initially used to balance the need for cash with the need to farm gradually devolved into longer and longer town stays.[46] A new urban-oriented living pattern ultimately created a housing dilemma that the companies determined to resolve in their favor.

The best illustration of this plan could be found in the strategy of Anglo American's Ernest Oppenheimer as employed during the 1930s and 1940s. Oppenheimer believed that the best means of securing a steady, submissive work force was to offer housing compounds with amenities of sufficiently high quality to make mining jobs both attractive and necessary. The partial model he had initially tried with mines in South Africa included dining-halls and dormitory rooms attentive to both the health and control of the local labor force.[47] Workers appeared to appreciate these facilities but South Africa's legally constructed racial divisions did not allow Oppenheimer adequate latitude to test his overriding aim. That aim was to provide enough opportunity for African miners to stay interested in working without the companies themselves relinquishing control over them. Northern Rhodesia, free from restrictive residential laws, was an excellent place for Oppenheimer to attempt his full-blown vision.[48] Such demonstrations of interest in worker welfare might further help diffuse tensions between the large African workforce and the potentially vulnerable, much smaller European managerial community. Focusing on Oppenheimer's principles, the private companies eventually began supplying cinder block housing for families, plumbing, electricity, and regular distribution of the staple food, mealie meal, in quantities and qualities the UK government simply could not match.[49] Over time, cinemas, libraries, schools and community centers also sprang up in the mining compounds along with endless organized sporting activities.

The Oppenheimer model worked in that it helped attract and maintain a labor force of the size required to keep the mines running. It also produced some consequences he might not have envisioned. For one, the very structures intended to keep workers docile seem to have enhanced their labor organizing abilities since the physical closeness of township dwellings allowed workers easy means by which they could spread information. For another, the competition that Oppenheimer's compound model engendered between European and African mine workers prompted improvements to European quarters. These, in turn, were countered by British government pressure to provide even more upgrades to the Africans' facilities. Eventually Oppenheimer complained that the burgeoning Copperbelt environment represented something closer to "a mining Utopia" than it did a mining camp.[50]

A third arguably unanticipated consequence of the Oppenheimer model was its endurance long after Zambians had taken public stances against colonial patronage. Even when the mines were nationalized in 1970 they kept in place policies and practices begun for the private corporations' benefit. Where, for example, Oppenheimer had valued control over the workforce via a sort of infantilizing infrastructure, Zambian mine ownership spoke of keeping workers contented by taking care of innumerable personal wants. Regardless of motive, the mine compound structure seems to have elicited a great emotional attachment. More than a decade after the entire system had begun to crumble through neglect as it was doing by the 1980s, and long after nationalization had ended as it did with the privatization process begun in the 1990s, what Zambian miners most frequently mentioned as the greatest loss associated with privatization was the townships' disintegration.

African Advancement and Federation

Yet, by the early 1950s Copperbelt miners were sufficiently organized and confident in their abilities to make their interest in taking on more responsible, better paying mining jobs a public sticking point for the companies. Many in the European worker's union were threatened by the prospect of Zambians taking over their jobs. The notion of "African Advancement," or efforts to train and employ Africans for work in positions that had previously been held by expatriates, was initially driven by Zambians themselves. But, unfortunately, by the time an official agreement had become corporate policy in 1955 it was probably too late to achieve a universally equitable and economically viable transition from expatriates to local employees. The gap between the general worker's educational level and the sophisticated technical skills needed to run an industry of such size and complexity was particularly huge. In retrospect, analysts with close knowledge of Zambia's mining history suggest that training Zambians to take over successfully from expatriates should have been an effort that started years earlier.

As the discussion of the 1935 riots should also have made clear, social tensions and political unrest accompanied the industrial mining endeavor from its inception. But formal, organized labor resistance was only one method that Africans employed to protect themselves. They also used their faiths and religious associations of which independent prophetic movements such as Watch Tower were notable examples. These more directly culturally connected associations could, according to some, critique the economic and political situation more stridently than could the African mission churches.[51]

The combination of labor-based initiatives such as African Advancement plus the socio-religious-cultural stirrings of Watch Tower and other movements caused European residents to become increasingly nervous at the prospect of a rising black African majority. This led to a revived interest in Cecil Rhodes's old vision of establishing an amalgamated regional power.[52] Years before, Northern Rhodesians had balked at the idea based on their dislike of possibly being dominated by a larger population of Southern Rhodesians whom they perceived as having different temperaments and objectives than their own. As the push for more African responsibility and power unfolded, though, many Northern Rhodesians voiced preference for the risk of southern white domination over the clear prospect of black majority rule. The copper mines finally beginning to prove profitable, Northern Rhodesians also felt that they would be in a better position to stand up for their own interests over against those of the Southern Rhodesians.[53] So Federation was proposed and then, in 1953, officially bound the two Rhodesias together along with Nyasaland (modern day Malawi).

Between proposal and implementation, however, were a wide-ranging series of stormy debates. Federation was publicly presented as an arrangement intended to benefit the entire region's development. In fact, one of the most intriguing aspects of this history is the way in which public discussions of Federation were framed in terms its being a "partnership" that well-intentioned Northern Rhodesians seemed to think offered the best means of avoiding either extremes of black or white nationalism.[54] There is also some evidence that Great Britain backed the plan in part to help forestall South Africa's white settler population from moving more aggressively northwards.[55] And it was the potential of such an "ideological frontier" thrusting itself beyond the Zambezi that most distressed Africans.[56]

In fact, Africans of the region were uniformly and vocally opposed not only to the idea of Federation but to the way in which the UK's policy position on the matter shifted dramatically over a period of only two years. In 1951 the UK appeared ready to accept that the majority of Northern Rhodesia's population disapproved of the scheme. But by 1953, Federation was basically announced as inevitable. According to some observers, this decision was all the more remarkable for the way in which Northern Rhodesians appeared to convince themselves that over riding the clear majority opinion on this matter was ethically acceptable. One European critic of the plan predicted that Africans would neither forgive them nor forget such a betrayal.[57]

Indeed, ultimately Federation resulted in just the sort of domination that both Northern Rhodesians and Africans had feared. While economic arrangements between northern and southern territories had always been lopsided, more money than ever before began flowing out of Northern Rhodesia to fund massive infrastructure development south of the Zambezi River.[58] The first such projects were railway transport systems and roadways intended for copper export. Later, a dam and power plant also intended for use by the mines and originally planned for near Kafue were relocated south to Kariba.

To Africans who had been against the amalgamation idea all along, the Kariba dam project particularly seemed to encapsulate all the reasons why Federation was such a bad idea. The highly publicized project entailed forcible removal of 30,000 people and thousands of animals from their lands in an effort famously celebrated as Operation Noah.[59] While hailed in the European context as being an example of cultural and ecological sensitivity in the face of modern progress such massive displacement was seen by Africans as simply a microscopic foreshadowing of what Federation would do for land rights issues throughout the region.[60]

Africans on both sides of the Zambezi River were deeply affected by the Kariba dam project, but those in Northern Rhodesia may have felt Federation's overall threat most keenly. Many had worked in Southern Rhodesia and were familiar with the territory's restrictive policies on land rights and trade union organizing.[61] In their minds, Federation would simply speed up the white settler's land grab, a particularly odious prospect given the differences they saw between the southern settler population and their northern European neighbors. In general, expatriates who came to work on the Copperbelt immigrated from mining communities in England, Scotland, and Wales and frequently returned there upon retirement.[62] Southern Rhodesia, by contrast, seemed to be increasingly settled by farmers descended from Dutch ancestors who had gone to South Africa about the same time that Puritans had traveled to America. Over the generations these Afrikaners, as they became known, developed large farms and tenacious beliefs about the innate superiority of white people. Africans subjected to Federation found that it actually mattered which European group would predominate. While they understood that neither group would have African interests primarily at heart there were still some relative differences. It was one thing, for example, to be treated as an inferior economic class by a shorter-term population that liked golf and lawn bowling and quite another to be thought racially inferior by a permanently settled group of well armed farmers.

Experiencing increased dominance of Southern Rhodesian interests over those of Northern Rhodesia's residents made Federation feel to Africans a double betrayal. Though they had never asked for colonial rule, some at least had come to view the British government as a benignly paternalistic presence. The process of devolution to black majority rule in Northern Rhodesia had been messy during the decades preceding Federation, but it had appeared inevitable. To be then forced into relations with a dominant region intent on keeping white minority rule, as Federation was doing, seemed a galling set back.[63]

Over time, however, it became apparent that African Advancement efforts already underway prior to Federation and the generally evolving public consciousness regarding Britain's colonial project itself would prove stronger than the alternative that Africans had feared.[64] Within half a dozen years Federation looked obviously untenable. During the Federation years, the Copperbelt experienced a number of "unpleasant incidents" as one observer described them. These included vandalism, arson, beatings and road blockades that were variously attributed to the pre-independence movement in general and the United National Independence Party (UNIP) in particular.[65] Within a decade of its implementation the arrangement was officially dissolved. Northern Rhodesia was back on course to become politically independent of Britain in 1964.

A number of impediments still confronted black majority rule, though. Many of these concerned the mines and everything seemed to carry political and economic repercussions. For example, by the time serious independence preparations were underway Northern Rhodesia had a flourishing economy and a cosmopolitan population. But the region's wealth was quite unevenly distributed. While on paper it appeared that Northern Rhodesia was in a strong economic position, in actuality the Copperbelt—where that wealth was generated—was something of an island of urban prosperity amid a sea of rural poverty.[66] Designing a system that would work for more equitable wealth distribution promised to be one of the post-independence government's most urgent and daunting tasks.

Compounding this problem was something not immediately obvious to the country's internal leadership or to outside observers. This was simply that for reasons having to do with social and physical distance much about how the colonial system really *worked* was never apparent to Africans. Throughout the colonial era, racial divisions had kept Africans from seeing what went on in the upper ranks of European power and control, much less from learning their thought systems and action plans. These same divisions generally kept well-meaning Europeans from getting to know Africans on more than superficial bases.[67] Additionally, since bureaucracies in Great Britain complemented or supplemented much managerial work being done at the local level it could appear as if relatively few people were needed to run Northern Rhodesia's quite complex society.[68] Further, the jobs that many expatriates occupied could, from the experiential distance of African observers, seem to consist of largely sitting in comfortable offices and talking. As Africans thought about independently governing themselves using a system they had not designed, it may well have looked to them much easier than the process turned out to be.[69]

There was also much about how colonial business was done in Northern Rhodesia that seems to have made Zambians less rather than more prepared to assume responsibility at independence. Lack of sufficient educational opportunities was a perpetual problem exacerbated by Federation. During that period in Zambia's history, European children's education was understood as a federal concern backed by considerable funding. African children's education, on the other hand, was left to the territories: Northern and Southern Rhodesia and Nyasaland. Consequently, Northern Rhodesia, already economically behind in comparison to its southern neighbor, spared few funds on African education and what was offered went largely to support the primary grades.[70] Post-secondary schooling was exceptionally difficult to come by in Northern Rhodesia since the nearest university was in Southern Rhodesia's capital, Salisbury (now Harare).[71] As a result, at independence in 1964 Zambia had only about one hundred university graduates (including one lawyer, one engineer, and two doctors) and one thousand secondary school graduates within a population of over three million.[72]

The low quality of Northern Rhodesia's African education system eventually affected the new country's governance and in ways that outside observers would likely not have predicted. One example can help illustrate. Zambia adopted a parliamentary style of oversight modeled off of a system that appeared to work well for Great Britain. But crucial differences inspired by historical, economic, even geographical circumstances meant that what worked in the UK was not a guaranteed success in Zambia.[73] So when Kenneth Kaunda's UNIP Party won a decisive majority of legislative seats in pre-independence elections, guidelines borrowed from the UK model dictated that the distribution of these seats be representative across the country. This meant in Zambia's case that forty-one of the fifty-five total seats had to come from rural constituencies. Yet for decades, Zambians had been moving into the urban areas where the country's few schools were located. So citizens possessing the education and skill levels needed to take over government jobs simply did not live in rural parts of the country. Many of those who ended up being appointed, then, came to their tasks with distinct educational disadvantages.[74] In essence, Zambia's government was, from the very beginning, structurally incapable of drawing in the highest caliber workers.

Independence, the Kaunda Era, and Nationalization

In the face of such disadvantages Kaunda, as Zambia's first president, generally appeared to hope that the country could create something new. This hope existed despite colonial failures and regardless of pre-emptive setbacks such as the educational level of members of parliament. In letters and essays produced during the early 1960s, Kaunda, a thoughtful and charismatic man, talked of offering the world a novel vision of a truly modern African state. Granted, the educational, managerial, and political grounding of the new country was very soft. But, thanks to copper, the economy would be strong.

So, with political independence inevitable, President Kaunda made a gesture toward economic independence in a famous encounter that took place only hours before the official handing over ceremony between the UK and the new government. Drawing upon extensive legal research and following a vigorous public relations campaign, Kaunda's administration offered the president of the BSAC a choice between £4 million net of tax to relinquish its mineral royalties or nothing. And he had eleven minutes to decide.[75] The case for wresting Zambia's mineral rights away from the colonial era companies was based upon reasoning that the main treaty the BSAC had initially used to enshrine the arrangement was made with the Lozi people. But the Copperbelt was Lambaland, not Lozi territory! Lewanika, the ruler with whom Rhodes had dealt so long ago, had never had jurisdiction over the Copperbelt. The BSAC acceded to the terms Kaunda offered,[76] giving Northern Rhodesia's African population additional cause for jubilation when they officially became Zambians at midnight on October 24, 1964.

Over the ensuing decades of his presidency Kenneth Kaunda proved an often contradictory but remarkably commanding force. As to his own preparation for office, he had only a couple of years of advanced education at Munali Training Center.[77] But he ultimately became an articulate speaker and prolific writer. While working for Zambian independence he appeared to have been less hated and feared by European settlers than was his neighbor, Robert Mugabe, and there were fewer opportunities to accuse him of leading organized violence.[78] Once in office, however, Kaunda appeared to be a demagogue to some Western observers while various Zambians criticized him for cultivating a deification of the presidency.[79] Despite the flair of his last minute showdown with the BSAC over mining rights and his extensive, thoughtful critiques of the economic principles upon which the colonials had established industrial mining in Northern Rhodesia, Kaunda's general sentiments regarding capitalism appeared murky. He was said, for example, to get along quite well personally with Harry Oppenheimer[80] and, in the early days of his presidency, seemed sympathetic to the idea of privatized mining as being the best way to manage the country's greatest asset.[81] Yet once the mines came under government control, mining management seemed collectively hostile especially toward Anglo American, their former principal shareholder.[82] Kaunda was never able fully to implement his dreams for the new country and, eventually, popular disillusionment with his socialist policies led to his ouster in 1991. By the mid-2000s, however, difficulties with privatization and the leadership that Zambia's first president exhibited in retirement regarding social and public health concerns[83] had produced a resurgence of public affection for the vigorously fit, elder statesman popularly known as "Super Ken" or simply "KK."

Varied perceptions of his presidency could lead to any number of historical judgments. But Kenneth Kaunda's legacy deserves nuanced and considered thought both for what he attempted to do and how he was constrained in doing it. For much of his presidency he was generally portrayed sympathetically by

Western media sources due to his political skill and personal charm.[84] Beyond most of his contemporaries and perhaps more than his successors Kaunda presented for Zambia and the world an overarching vision of an independent African nation-state based upon two interrelated principles. The first was a sympathetic stance toward the socialist values of Eastern bloc countries over against the capitalist West. The second was a Humanist philosophy that, when infused into governance, was envisioned as helping make Zambian society more like traditional African societies.[85]

In regard to the first principle, Kaunda acted similarly to many other first generation African presidents. During the 1950s and 1960s Zambia, like most of Africa, was subjected to various bids for loyalty that were closely connected to the super powers' Cold War standoff.[86] Because of his experience with Western-based interventions, Kaunda ultimately decided that friendship and resource sharing with the USSR and China were preferable to economic development strategies practiced by the colonial powers. Needing money and expertise for Zambia's own development plans, its new president turned East rather than West.[87]

The second principle on which Kaunda built his governing philosophy may have exhibited too many contradictory values to prove as fruitful as hoped. While his philosophy was constantly featured in the mass media and formal educational institutions, the extent to which it was actually endorsed or accepted at a grassroots level was never clear.[88] So, crucially, and somewhat counter intuitively, ordinary citizens were only minimally involved in deciding how modern Humanism and socialism actually related to traditional cultural values.[89]

Additionally, when linking Humanism with socialism, Kaunda appeared to send mixed messages about what he meant by the latter term. Sometimes he spoke in theologically-laced idioms evocative of the Christian socialist tradition of early industrial era Anglican Britain and sometimes in economically-based idioms more applicable for the theories of Karl Marx. Within a matter of several years (and some have argued with justifiable cause given his further interactions with Great Britain)[90] the Marxist vantage point appeared to predominate as his links to the Soviet Union and China strengthened. This shift eventually helped prompt local church leaders, who had initially been quite supportive of Kaunda's governing philosophy, to issue strong statements calling for public discussions containing alternative viewpoints based upon their own readings of Biblical principles.[91] And at least in later life, Kaunda himself talked mainly of a clear compatibility between Christianity and what he had tried to do with Humanism.

> [T]he Humanism we are talking about is something which accepts human beings as a fellow human being regardless of anything artificial. And . . . when we were given the opportunity to lead ourselves—to rule ourselves—we came to analyze this this way: . . . All of us are the same nation regardless of 70 different tribes. . . . So it's "One Zambia, One Nation." We all are one nation. . . . [S]ome might say it's religious, well we're talking about God so it must be religious but it's not something that was saying we were going to be Humanists and—but not Christians. Or we're going to be Humanists but not Hindus, you

> know. . . . Capitalism as we see it was to build wealth—much wealth—for a handful. What love is there for your neighbor if you're going to exploit him or her to the point where he is struggling? She is struggling and you say, "That's all right," you accept that? I believe that that love of your neighbor simply means that love. And if you love your colleagues, you love your brother or your sister, you can't at the same time go against Christ's teaching: "love your neighbor."[92]

Some contextual consideration is important at this point in the history as well, for by the 1960s, the secularization thesis[93] had taken hold solidly in professional development discourse and in relations generally between the West and the rest of the world. This had profound theological and policy implications for what could be understood as the public role of religion generally.[94] What it also implied for Zambia's case was a quite nuanced and complex interrelationship among external players, internal political actors, the churches, and traditional cultural understandings. Any outside attempts at assessing the situation must be done with cautious respect for major unknowns created by this complexity. What was said or done and why, for example, should not be interpreted as reflecting the sort of dualistic understanding prominent within America at the time that associated the West with democracy and Christianity and the East with communism and atheism. Nor, for that matter, should it be assumed by outsiders that the contested relationship that developed between Kaunda and the Christian churches of Zambia was an argument entirely conducted by independent, internal equals. The extent to which a private discourse existed that represented its own dynamic mixture of Western influences, African cultural values, hybrid spirit movements, bald power struggles and the like will always remain a mystery to outside analysts. But the key point to maintain is caution about ascribing exacting descriptors to situations such as those encountered by Zambians in the early independence years.

Indeed, exacerbating internal constraints on Kaunda's broad vision for governance, external events hampered rapid transformation of Zambian society. The most pressing of these proved to be events surrounding the relationship with Zambia's closest neighbor. Southern Rhodesia, scheduled to attain political independence around the same time as Northern Rhodesia and Nyasaland, took a radically different path. The region's European minority population, realizing that independence would entail their living in a majority ruled black nation departed from the agenda set for them by the UK. Shortly after Zambia's one-year anniversary, Southern Rhodesia's white settler government announced a Unilateral Declaration of Independence (UDI) from Great Britain and instituted plans for a permanently white-ruled country known simply as Rhodesia.[95] The announcement precipitated an almost immediate and lengthy civil war. Britain backed the African nationalist freedom fighters, including the group headed by Robert Mugabe, in their opposition to the regime of Prime Minister, Ian Smith.

The trauma of Zimbabwe's civil war is still evident there decades later and, for Zambia, the conflict had a peculiar, lingering devastation. Operating under an official state of emergency for years, the country harbored insurgents, brokered talks aimed at ending the war, and suffered major damage to roads and bridges used for African troop movements but bombed by the Rhodesian Air Force.[96] Much of this latter hardship resulted from South African and Southern Rhodesian sympathizers who infiltrated Zambia and attempted to disrupt coordination among various African groups.[97] Additionally, the old colonial era rail lines used for copper export and routed through Southern Rhodesia were declared off limits. So the mining companies had to find circuitous pathways out of Zambia to Mozambique's port.[98] Lack of suitable transport through Southern Rhodesia had importation implications as well. Most crucially, Zambia was forced to airlift in fuel until, with the aid of China, a pipeline eventually could be constructed to Dar-es-Salaam.[99] Aggravating this situation was a constant worry that the mines' power source, located on the southern side of the Zambezi River, could be switched off by the Rhodesian government at any moment. So a new power plant had to be constructed on the north side of the Kariba dam.[100]

At the outset of the UDI period, Britain sought Zambia's help in isolating the Smith regime economically.[101] Zambia attempted to comply but the two countries cooperated in a strained fashion. There were several reasons for this. For one, Kaunda's resolute stance in support of African liberation movements throughout sub-Saharan Africa put him out of favor with many Western interests. As the Zambian government's loyalties became increasingly obvious, many European expatriates who had stayed on after independence now began to leave. For their part, many Zambians felt a certain justifiable disappointment in the way the British government conducted its policies regarding cooperation against Ian Smith. Kaunda's compliance with British sanctions against Rhodesia had come about because the UK had guaranteed compensation to Zambia in return. Such support would include financing for Zambia to build alternate transport routes and power supplies. But two years into the conflict all such funding was suspended.[102] There are conflicting accounts as to why this happened, but the overriding point was that Zambia's early years were made vastly more difficult in part by situations far beyond the control of its nascent government.

All the while, the mines continued producing copper but the country's problems created additional difficulties for the Copperbelt Province. As government tried to manage an array of crises and concerns, the lack of a sufficiently educated and trained workforce now became acutely apparent. The only place that did have a modicum of well-trained talent was the mining community, so the mines were the first place that government looked any time a managerial need arose elsewhere.[103] This pool was not, however, sufficiently deep enough to meet demand and it became even shallower as positions vacated by departing expatriates increased in number. Simultaneously there was an urgent need for reinvestment in mine infrastructure. Kaunda's government had already accused the colonials of not having paid sufficient attention to this matter.[104] But, under his leadership neither significant reinvestment nor exploration for new mineral deposits went forward. It also appeared as if mine accidents were on the increase. An independent investigator assigned to look into the matter said lax discipline and deteriorating standards seemed at least partially to blame.[105]

Part of the reason for this lack of initiative on the part of the Kaunda government could have stemmed from public pressures to do other things. The people's expectations for rapid completion of innumerable development projects from road construction to schools and hospitals were quite well known.[106] To pay for such projects the government had to rely on mine revenue because it was the country's only substantive income stream at the time. At first, the draw rate on revenue was fairly painless. Production was good and the world copper price was high thanks, in part, to the United States' involvement in Vietnam. Over time, however, Zambians noticed a continual drain on the mines' finances but no corresponding improvement in the country's infrastructure. Kaunda's vision for creating a new African society was failing to deliver upon the peoples' hopes. The state was not creating jobs as anticipated and private investors were not coming to Zambia as forecast.[107] Worse yet, some of the investors already present in Zambia before independence began displaying sympathy for the Smith regime.[108] In retrospect and given this constellation of needs, interests, and constraints, Kaunda's move to take full economic and managerial control of the mines does not seem a particularly surprising action. But at the time, his government had been giving what might be thought of as mixed signals about long term intentions toward the mines. By some accounts, Kaunda had repeatedly assured the public and mine executives that the government would not take over the mines, but nationalization of a number of other enterprises was announced at a UNIP conference at Mulungushi in early 1968.[109] In late 1969, he invited the mining companies to sell 51% of their shares to government effective January 1st of the following year.

The announcement came within a much longer speech to the UNIP National Council at Matero Hall, Lusaka. Looking back, some observers felt the announcement was such a surprise that it overshadowed other things Kaunda said about Zambian workplaces. He said, for instance, that there would be a complete wage freeze and that union strikes were banned.[110] Other observers were more prepared for nationalization. Perhaps, some have suggested, the reason the mines had been distributing annual profits as dividends all along was in anticipation of eventually being taken over by the state.

Following on Kaunda's announcement, a government pamphlet asserted that the nationalization plan really flowed from the original colonial era mineral concessions.

> The Constitution granted to us at the time of Independence by the British government was so designed that it denied power to our government to act against such concessions and companies . . . [But the President] has announced he is abolishing all these concessions. And the rights to mine the minerals in our Zambia come back to where they belong: to us today; to our children tomorrow; and their children the day after—and so it will go on for all time. Never again will these rights to our minerals belong to any foreign companies or any individual.[111]

At the time of the announcement, Kaunda declared that nationalization would be done in such a way as to continue allowing significant capitalist involvement in mining. Government would become the majority shareholder, but various tax and foreign exchange incentives would keep the private companies interested in remaining as investors. Apparently unwilling to jeopardize the country's economic base just to make a political statement the government carried on negotiations with the mining companies for several months.[112]

Ultimately, however, the program extended far beyond a 51% share ownership in the mines. Many other private companies were taken over and a wholesale restructuring of Zambia's mining industry reduced private investment as well. State corporations serving as holding companies were formed and the chairmanship given to the President himself. Name changes for the largest mines also signaled a shift toward new, local interests. Roan Selection Trust became Roan Consolidated Mines, Ltd. (RCM) and the Anglo American group became Nchanga Consolidated Copper Mines, Ltd. (NCCM).[113] Such shifts were not simply symbolic. The Kaunda government also reworked corporate tax legislation in its favor and managerial changes followed the mines' name changes. With the government's subsequent termination of contracts with pre-independence era investors, for the first time in their histories Zambia's industrial mines became self-managing companies with Zambian directors.[114]

All these changes took place within the context of a formal program in which the presence of expatriate influences was scheduled to give way to that of Zambians themselves. The program, known as Zambianization, was initiated only two years after independence.[115] Kaunda's government described it as a way of reducing the country's dependence on imported skills and promoting Zambian interests. Tied in spirit to the African Advancement program of the 1950s Zambianization was in full force by the mid-1970s.

At least two major operational problems hampered the program's success, however. One was simply the monumental nature of moving Zambians into jobs formerly occupied by expatriates. By industry accounts, the logistics of coordinating such a shift would need to involve over 60,000 people in more than 3,000 job categories.[116] There was also the dilemma posed by trying to fit massive numbers of Zambians into jobs occupied by a comparatively small number of expatriates. After independence, the numbers of expatriates filling prime skilled and semi-skilled positions within the mines had shrunk to about 7,800. Making rapid progress in the transition from the group of expatriates to the larger population of local workers was accomplished first by moving Zambians into positions of authority within mine personnel departments and then fragmenting other jobs. In this way, two or three Zambians came to assume duties that had previously been assigned to one European. While this procedure definitely sped up the transition it also contributed to the eventual bloating of the mining company's workforce. Compounding the difficulties of this change were the educational capacities of a workforce competing for newly opened jobs. Available Zambians numbered about 41,000, but perhaps as many as 20% of these workers had only attained a primary school education.[117]

At the same time, Zambianization fueled the old 1950s resistance to African Advancement that European line workers had expressed at the time.[118] Many expatriates were disinclined to help Zambians over any knowledge and information gaps since the end result of such assistance meant the phasing out of their own jobs.[119]

The second major mechanistic difficulty with Zambianization was related to this last point and particularly concerned skills training for mining jobs. The mines had historically conducted training programs, but finding sufficient numbers of eligible secondary school graduates who could take on highly specialized jobs was an ongoing problem.[120] Sometimes recruits chosen were not the best candidates.[121] Often times, training program graduates were quickly recruited, or "poached," by other companies. So in all, quite a few beneficiaries of the mines' educational efforts never really put their knowledge to work on behalf of their sponsors. By the early 1970s, outside resistance was joined by an inside apathy that, according to an independent report, threatened the entire Zambianization process. In an ironic twist, the report claimed that apathy to Zambianization was coming from African trade unions and the government. This meant that the government itself was at risk of not honoring its own commitments to the people.[122]

As the outward flow of expatriates continued, those non-Zambians who remained were eventually found mostly in financial and technical positions.[123] Managerial jobs requiring overarching knowledge of the industry, vision and strategic planning frequently went to Zambians who demonstrated loyalty to the government regardless of their training or experience.[124] Ultimately, the idea of Zambianization may have been more optimistic than the results could possibly have delivered under the circumstances.[125]

Enter the Multi-laterals

Reasonable critiques of the Zambian government's policies aside, however, it behooves us to remain aware of larger forces that compounded the country's governance problems. In particular, a word should be said about the intervention of external economic interests specifically as manifested by the Bretton Woods multi-lateral institutions. At the time of nationalization, the country already had accumulated a modest amount of external debt largely from taking on bank loans for capital projects and hiring outside consultants to assist with managerial needs.[126] As the mines' income fell farther and farther short of meeting Zambia's development needs, though, government increased its borrowing. Rather suddenly, what appeared as a manageable debt load in 1969 became a debt burden by the early 1970s.

In fairness, not all of the debt was incurred because Zambia simply lived beyond its means. In fact, the rather sudden shift in the country's indebtedness was—at least in part—worsened by a massive cave-in at the Mufulira Mine in September, 1970. Thereafter referred to by the populace as the Mufulira Disaster, this was not the seminal event in Zambia's economic deterioration but it did precipitate a massive production drop during the mine's closure and rehabilitation.[127] Unfortunately, at roughly the same time as the Mufulira Disaster the world copper price dropped significantly. Zambia began struggling to meet obligations to the World Bank and turned for assistance to the International Monetary Fund (IMF).[128] A one-year agreement negotiated in 1973 offered temporary help with Zambia's debt obligations to the World Bank but the IMF arranged this in exchange for the country's compliance with a few economic conditions. On the list of conditionalities was the elimination of price controls over certain goods and services.[129] The government's administration of price controls prior to that time had been somewhat haphazard while the underlying rationale for them remained opaque.[130] Whether Kaunda's government employed price controls as part of a socialist ideology, as a calculated means by which to purchase citizen loyalty, or as something else altogether is still debatable. But, the popularity of price control measures was unquestionable, so their elimination became a far from trivial policy matter to the general populace.

Zambia in the 1970s subsequently experienced increasing chaos as living standards deteriorated and international debt accumulated. Less income meant less financing for maintenance and improvement of the Copperbelt's social infrastructure.[131] In the townships, streetlights and telephones were vandalized and the industry's local newspaper reported an increase in general social problems. Government publicized several interventions such as reducing copper sales and shutting overly costly sections of the mines. But the industry was now obviously suffering from what President Kaunda acknowledged as a combination of management problems and generally difficult economic times.[132]

Mounting chaos within the mining community mirrored turmoil within government. The one-party democracy Kaunda had introduced in 1972 was supposed to align Zambia more closely with African traditional values. But it also allowed the president to consolidate power. Within his cabinet there was internal dissention and Kaunda frequently reshuffled ministers' assignments. Ordinary citizens had already begun registering their disapproval by dropping out of the voting process.[133]

The ZCCM Era

By some accounts, the next major consolidation of the mines precipitated the industry's virtual collapse. In 1981, government announced the merger of the two self-managing companies (RCM and NCCM) into one conglomerate—Zambia Consolidated Copper Mines (ZCCM). Presented as an efficiency move, this consolidation was further said to be the logical conclusion to the mines' initial nationalization.[134] To those in the industry as well as outside observers, RCM was better managed and reasonably profitable at the time. But NCCM predominated in the merger even though its operations were "limping."[135] NCCM's Chief Executive Officer took over running the new, huge corporation, and brought with him the former company's hierarchical and formal work style.[136] Even though ZCCM's new logo combined the old Roan Antelope and Nchanga's copper alchemical sign, Zambians joked the only thing consolidation changed was that the N in NCCM had fallen over on its side.

The newly created company was huge and to the people, "Mother ZCCM" as it was sometimes known, was expected to provide employment, continue the social programs, and infrastructure maintenance that previous mine ownerships had begun. At its most expansive, ZCCM ran timber companies and farms and was responsible for managing much of the country's overall operations.

> We went into construction, right? We went into construction including elevators. We went into lodges. We started running the Kasaba Bay up in the North by Lake Tanganyika. . . . And all those resources now were coming . . . the managing director of the bus company came from within the company. The engineers were taken out from the company . . . and then it got to a position where if the government had a problem procuring medicines because we had a procurement agency in London, in the UK, they would then ask us, "Can you buy medicines for government offices because we haven't got medicine. . . ."
>
> When the Pope came to visit Zambia, he had a mass in Kitwe. We did everything. We cleared land. This was a bush. We cleared, put up a platform for him, all the things and so forth. Now the vice-chairman of ZCCM was in charge because he was a Catholic anyway. He was in the Church. But because there was nobody—you would look to nobody else. So this was a grand thing, successful, run by ZCCM. So we did this thing.
>
> When there was a mess in the football association of Zambia, you know, the whole of Zambia football association, they picked a guy from ZCCM. "Go and be the president." Somebody with influence to be the president. And therefore the secretariat would be somebody from ZCCM. So all that provided by ZCCM. And that's how the resources became so extended that there wasn't much now going back into developing the mines, you see, because of all that. We, basically, we were running this country.[137]

Consolidation did not, however, solve Zambia's economic problems. Strapped for cash, government began negotiating a first Structural Adjustment Program (SAP) with the World Bank and outside consultants began arriving in significant numbers.[138] Technical experts examined all aspects of the mines' operations, wrote numerous reports, and made recommendations, some ultimately spending the equivalent of five or six years on the Copperbelt.[139]

Policy measures accompanying Zambia's new SAP were those that typified such arrangements at the time. These included abolition of certain import restrictions, liberalization of interest rates, restrictions on government expenditures, and further removal of price controls. To this last point, however, the Kaunda government objected for why should Zambia drop its mealie meal and petrol subsidies on orders from countries that controlled the prices of their own farm goods and petrol production?

> [A]t that time they were making us reduce assistance. . . . And we reduced them[140] until we came to about three: it was mealie meal, fertilizer, and oil—fuel for transport and so on and so forth. And the argument then—they insisted that we should just scrap all of them. And we said, "Look at the list—what we have done. We have reduced—accommodated you up to—but mealie meal is critical to the needs of our people. We cannot do that. Fuel affects—the price of fuel affects the whole economy because transport and all the costs of production are going to go up if the price of fuel goes up. Fertilizer is agriculture. We want to promote agriculture so that our people can continue to sustain their

> production. . . . We've gone as far as we can go. . . ." Subsidies are used in America very heavily. Subsidies are used in France. . . . So in a way they were asking to do what they themselves were [not] doing in one way or the other. . . . It was not a question of simply saying "We don't want to do it," because we had been working with the IMF and the World Bank all along and supporting the programs and undertaking the programs that they were recommending us.[141]

Confusion and conflict erupted among policy makers and the public. The World Bank gave Zambia mixed signals on how to handle its monetary policy.[142] And, when government did finally do away with mealie meal subsidies, riots broke out on the Copperbelt.[143]

In 1987, government attempted to regain control of the country's economic policies. Kaunda abandoned the World Bank's SAPs and introduced his own program for the country's development.[144] Whether or not his strategy would ultimately have been successful will never be known, for his actions prompted bi-lateral donors to freeze their development aid and the country's foreign exchange reserves subsequently dwindled.[145] Kaunda's administration responded in turn with a ban on the importation of luxury goods, causing shelves in local shops to go bare.

Within two years' time, living conditions had worsened to such an extent that government was forced back to the World Bank and the reintroduction of SAPs.[146] By this time, however, the influence of periodic Copperbelt food riots had spread to Lusaka and, in 1990, there was an attempted coup. Civil society leaders, particularly those from the Catholic Church, stridently called for open, multi-party elections since civil war looked imminent.

> [T]he first attempt at a coup was that the people in the middle of the night heard about it on the radio and they ran out in the streets and they were dancing. . . . They'd had enough. I'm told that Kaunda wanted to fight it and I believe it that he was going to go hard—I mean really clamp down. And people and the churches were against him—all the churches combined together on the issue—I mean you used to get political diatribes at the pulpit in the Catholic Church every Sunday as it went on. The Catholic guys were pretty strident. Very strident. And they politicized their sermons overtly. And . . . somehow they, the combined churches, came together with KK and some other advisors and said, "Look, you've got a choice. Let this go through and let there be multi-party elections or civil war. Now what do you want? Because they will fight. We've reached the end now; there's no going back. You either let this go or you're going to have a blood bath. There'll be civil war." . . . [I]n fairness Kaunda had a choice and he let it go through. ZCCM spent a lot of money helping UNIP and Kaunda try to win the election. But it was too late then. People took the money and voted the other way.[147]

In 1990 and 1991, the government enacted constitutional changes that ended one party rule, enlarged the National Assembly, established an electoral commission, and allowed presidential candidates to stand from other than the ruling party.

The Chiluba Era and Re-privatization

Prior to 1990, the one party system Kenneth Kaunda had created effectively prevented organizing by opposition parties. So there was very little time to mount a rigorous opposition. In haste, a group of church leaders, disaffected UNIP party members, intellectuals, and business persons cobbled together the Movement for Multi-party Democracy (MMD), naming trade union organizer Frederick Chiluba as its head. MMD principally blamed Kaunda and UNIP for the state of the economy and the popular vote concurred with this assessment. An election held in 1991 brought Chiluba into power as Zambia's second president.

Frederick Chiluba's leadership record, like that of his predecessor's, is still being assessed by historians and political analysts. Ultimate judgments will most likely range widely. Many outsiders have already pronounced Chiluba to be the political leader who began restoring Zambia's economic well being after decades of unhealthy policy actions. In this view, whether Chiluba himself was a visionary or simply happened to be well positioned at the right time makes no difference.[148] Rather, what happened under his presidency is the important part. Many Zambians, in retrospect, have seemed to regard him rather well despite overall distaste for the privatization process he so enthusiastically embraced. Under Chiluba, the promise of a widely and materially prosperous society appeared at least plausible again as it had in the early independence days. No doubt some of the tactics Chiluba employed—such as selling miners' houses to them and declaring Zambia a Christian nation—helped bolster public perceptions about his presidency as leading the country into a new, self-reliant and more confident era.[149] Indeed, the early period of Chiluba's presidency was a hopeful time. Coinciding with Zambia's change in government the World Bank and IMF restored normal relations. Both the world copper price and the mines' production went up for a time and many new ideas for governance and business enterprise appeared in the offing. Local shops once again stocked their shelves and investors from South Africa and elsewhere returned.

It was, however, a short lived time of optimism. After a couple of bad years of copper production, the World Bank intensified interventions in the industry's affairs. This time the Bank's involvement reached into the operational as well as the more conceptual levels of planning and financing.

> World Bank used to underwrite loans to ZCCM in many cases, especially if we were borrowing from the Western world. So, whenever they did that, they used to come regularly like once a year or once every other—they would bring out a technical team to audit the mines. . . . [In 1994] they visited each mine. They spoke to the general managers. They spoke to management. In some cases, they went underground. They looked at performance of the division. But they went and made a decision that now, because now we were looking for money from them and all that, they were going to be instrumental in lending us money. So they said, "If you want money from us you've got to make these changes. So the general manager at Nchanga, the general manager at Nkana, should be replaced as soon as possible." And the Chief Executive had very little time—

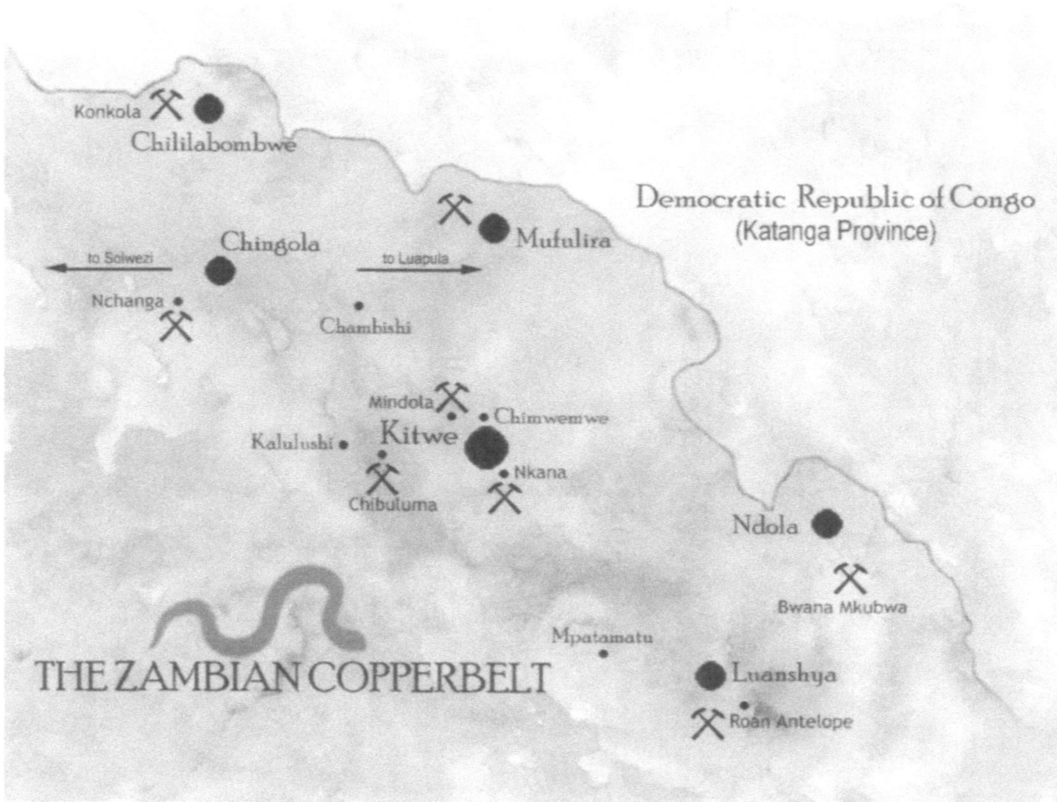

Map of the Copperbelt Province. © Sarah Louise Sears

Chingola open pit mine. Photo by author.

A donafish rendering, Chimwemwe, 2004. Photo by author.

Monument to colonial era copper discovery in Mufulira, 2004. Photo by author.

Copperbelt township street, 2004. Photo by author.

probably only had a month. And the best therefore that he could do was get an expatriate who was a consulting mining engineer . . . and took him to Nchanga. He went for a guy that had retired from ZCCM—an expatriate. Pulled him from retirement and made him general manager [at Nkana]. To suit the World Bank requirements. And, upon that, the World Bank released the funds.[150]

The Bank also extended its influence throughout the country, implementing procedures it typically employed throughout the developing world and in accordance with the ideology of development theory during that era.[151] The Bank's main policy prescription for Zambia was a three-pronged "strategy for renewed growth"[152] that included trade liberalization; economic diversification particularly through agricultural development; and privatization of the country's nationalized industries.[153]

From a purely economic standpoint, privatization was not the only option for Zambia but it was the most viable. Commercialization could be tried. But this would involve massive refitting of the mines' infrastructure. Since government had 60% of the shares but no money where would funds be found for this option? Liquidation was also a possibility. But that would mean eliminating over 50,000 jobs—equivalent to disbanding the nation's army.[154] How could the results of such a move be borne either economically or politically? Privatization, on the other hand—though it would involve selling controlling interests in over 200 state-owned enterprises ranging from travel companies to drilling firms to stores, banks, and utilities—was not a wholly new concept in the nation's history. In fact, Chiluba could speak of privatization as simply a return to familiar, pre-independence territory.

There was no such thing as ZCCM in 1964 and before. You had Roan Selection Trust controlling a certain number of mines including Luanshya and Mufulira. And then you had Anglo American running Nchanga, Konkola, and Rokana. Then you had, I cannot remember how many private companies which were working for the mines, supplying to the mines, and contracting to help the mines in one way or the other. Very many around the Copperbelt. And the Copperbelt was the center of industrial activity in Zambia. And so we didn't have a command economy at the start of it all. And at the peak of economic boom in Zambia it was a private economy which was running. So the private sector was there even before independence.

When we got into independence, there was this one very common approach by African states after defeating colonialism we believed strongly and we thought it was not coming, but it has come, after defeating colonialists we knew that they were no longer going to come with their colonial heads in politics. They would come as neo-colonialists trying to control us by using resources. Our own resources. And we have seen it today. What Nkrumah[155] said in 1958 has come true today. So every African government, not only in Southern Africa but . . . the majority of them felt the best way to fight colonialists was not only to win political independence but also turn the economy from the market to the office. From being market economies to being centrally planned economies where government could control the planning every single day. That way we could say, "Let's stop that bus which is run by Western imperialists and only

> set up a company run by comrades who were freedom fighters. We are sure they will not overcharge the passengers." . . .
>
> And so here we tried to swing because we were reacting to imperialist and colonial forces that had oppressed us. So we tried to swing not only politically but also economically by introducing economic systems which were contrary to what the West believed. And—but we over did it. We over nationalized. We went up to 80% of the economy totally nationalized. And the results were catastrophic. The results were catastrophic. So when you talk to me about privatization, it is not as if we were dreaming about something that had never happened. In fact returning to the status quo before independence. The status quo that proved extremely successful.[156]

Of the three possible restructuring strategies available to the nation, privatization was the model obviously favored by the World Bank. Official external analyses of the country's ills could be scolding in tone with the Bank and IMF particularly faulting Zambia's leaders for their lack of economic diversification and for over dependence on copper.[157] The Bank criticized Zambian leaders for squandering economic diversification opportunities; failing to read market signals; and being generally too single minded about copper production.[158] Still, there must have been something of value to all those "wasting assets" on the Copperbelt, for out of Zambia's entire privatization process, the mines represented, "the biggest prize."[159]

No mention was made in World Bank documents of an historical context that included Rhodes's own single minded interest in the region; nor of how the BSAC's own administrative deficiencies were made up for by the UK; nor of the critical educational and experiential disadvantage with which Zambia had begun life as a political entity. From a strictly economic standpoint, these factors might have had no bearing on the country's present and future prospects. But there is a certain philosophical irony in Zambia being criticized for concentrating on copper production when, in so far as the original colonial vision was concerned, copper production was the region's *raison d'être*.

This history makes it all the more remarkable that many Zambians somehow took to heart an identity prescribed for them from outside. Over the generations, many had apparently become reconciled to modern, industrial mining. In fact, the Zambian public had managed to turn the existence of imposed industrial mines into a deeply felt sense of national pride.[160] Their roles in support of that industry formed part of a collective being much richer than externally conceived policy making could reflect. Consequently, the forced sale of the country's huge range of nationalized companies—with the mines being the crown jewels—engendered years of "raging controversy."[161]

This controversy had a dramatic effect on the mines' operations even before anything was put up for sale. ZCCM had already begun trimming its huge workforce a few years earlier, sending many experienced workers off with severance packages.[162] Left behind were less experienced workers who carried on in an atmosphere dominated by uncertainty and worry. The mines were also soon in arrears to private supply contractors and their daily losses were said to be fantastic amounts of money.[163] But these figures came from those most interested in the mines' dissolution. So could the numbers really be trusted?[164] It was clear that some level of production had to continue, but how to do this in a context of increasing scarcity? Company personnel found themselves reusing spare parts and straining credit with local suppliers in order to maintain production.

> [T]he people in production engineering . . . went back into short cuts. Washing a used bearing . . . things that they had thrown away to the salvage yard. You would go back and see how we could clean it up to fit in because there was no bearing on the shelf in the stores . . . So they started to realize they were taking short cuts: . . . "Let's keep these mines going; let's keep these things—let's keep some production going." So they realized that we were short of funds. . . . It was like nationalism; it was like a national duty. Because even our—our salaries including myself—were not as good as we are being paid here under privatization. But we just said, "This is our country. We cannot afford to see these things sink."[165]

As the privatization program continued to unfold, the majority of Zambia's nationalized companies were sold before ZCCM itself was considered for sale. Given that the participants involved had thus compiled some history with the process it might be assumed that the mines' sales proceeded fairly smoothly. Yet, when it came time to consider the mines, bickering among analysts and others over how the sale should be structured delayed negotiations for two years.[166] Mining experts and chemists advised keeping ZCCM intact. They reasoned this way because the ore bodies of the Copperbelt are sufficiently different within the space of even 50 kilometers that they require specialized mixing of concentrates. If the huge company were broken up into smaller units, who could say whether there would be adequate cooperation across new company lines to ensure that such production details worked? Financiers, however, advised "unbundling" or dividing ZCCM's assets into pieces.[167] Unbundling the conglomerate, they said, would allow individual companies to control the entire copper production process from underground development through smelting and refining. It would also make clear which assets were core business concerns and which were not.[168]

This last point was a significant matter since, by the mid-1990s, the mines' non-core assets were considerable. Most obvious among these were the townships where miners and their families lived. What would become of these areas was of enormous political and economic import. Massive worker retrenchments could be expected no matter how the mines were sold, so the Chiluba government insisted that the homes ZCCM owned be sold to their occupants. He reasoned publicly that turning the township dwellings over to the miners themselves would provide them with some economic stability in the face of anticipated job losses.

> [T]hey talked about mass redundancies which were on the way; coming up. And these people were going to be declared redundant. Where on earth were they going? They had stayed in those houses for many years. They had brought up their families in those houses. Their families have been going to school in these areas and not anywhere else. But if they were declared redundant, they

> were going to lose their houses. They were going to be destitute. And I decided never to allow that to happen because these are the people that built those houses for many years. And so I said we are going to take away all the non-performing assets, reduce the balance sheet by taking away the non-performing assets and leave only plant and equipment. Plant, machinery, and equipment to sell to them—which included houses. And we also left a small portion of a number of houses, maybe between 50 and 100, for management. But the rest of them we sold—I don't know—what price you call it. It's not market price. . . . It's not giveaway price. It's—I don't know what price but it was a wonderful price which everybody afforded. And so today we don't have any war to fight. People have been declared redundant but they have turned elsewhere. As long as they have their shelter, they have a roof. They have just turned elsewhere and begin to do some business on their own. And they live and survive.[169]

Chiluba's critics have noted that he also benefited enormously by this policy since it was understandably quite popular with the people.

Eventually, it was decided that ZCCM would be unbundled, but this decision did not end the sale controversy. Disagreement simply shifted to questions over who should purchase which parts of the business. The privatization process may have reached its dramatic high point in the late 1990s during negotiations with a consortium put together by several multi-national mining houses. By some accounts, the Kafue Consortium was interested primarily in ZCCM's Nkana, Nchanga, and Chambishi assets.[170] By other accounts, Kafue Consortium wanted the entire company, something Chiluba insisted he would never accept.

> Had we accepted Kafue Consortium we would have made shifting ZCCM as one unit—a public monopoly—into Kafue Consortium—one unit. But this time a very dangerous private monopoly. And as a trade unionist I could not accept that. I could not. So my government stood firm.[171]

Discussions had reached a very late stage in London when a last minute disagreement caused the Consortium to walk away. The official version was that the $400 million bid was too low for the Zambian people. Private speculation was that perhaps some bribes were expected. Yet another possibility was simply confusion over the point at which copper becomes meaningful.

> [S]omebody sat down and calculated what they thought they would get by privatizing. And they thought they were going to get millions of dollars because they—I think the guy that did it calculated the sales revenue if you got a certain tonnage of copper over a certain number of years. And when the companies came in and started offering them 70-80 million, because that's all it was worth to them, there was such a shock. And then recoil, "These people are out to gouge us again. They haven't stopped. You know it's just the eagle with the talons again ripping us off big time."[172]

With the Kafue Consortium deal collapse, talk turned to questions of other buyers. Should old familiar investors, such as Anglo American, be invited to resume more significant roles? Anglo American had all along retained a minority share ownership in the mines. It was also becoming increasingly involved with Konkola Deep Mining Project (KDMP), casually known as Konkola Deeps, not far from Chingola.[173] The company's attachments to Zambia came primarily from Harry Oppenheimer and he was now elderly, but pushed the Company's board to return.[174] What about management buyouts? Some ZCCM executives attempted to put together plans.[175] Should new foreign sources be sought and, if so, who should they be? There was talk of diversifying mine ownership around the globe. The government appeared to think that inviting investors from various parts of the world to purchase individual mines would help bi-lateral relations. What about the Zambian government's and people's interests? Finance minister, Ronald Penza, assured the public that government would retain a "golden share" in each privatized package. This would give government limited special rights concerning such things as changes of ownership and mineral asset disposal.[176] As to the people, government would eventually float its shares on the newly established Lusaka Stock Exchange (LuSE). This, said industry reports, would "enable Zambian participation in the ownership of the new company."[177]

More vacillation and uncertainty delayed the process and now the world copper price dropped. Zambia then found itself being "a very, very willing seller" looking for "a very reluctant buyer."[178] But, since the World Bank and IMF were refusing further funding, the privatization program went forward.[179]

Throughout the entire privatization process, government did seek to balance mine ownership by inviting investors from different parts of the world. This effort was reflected in the range of buyers who eventually purchased portions of the unbundled ZCCM. In 1997, the first sale was completed. Kansanshi mine, the site of some of Africa's oldest copper mining and smelting activities, went to Cyprus Amax based in the USA.[180] The famous Roan Antelope Mining Company and nearby Baluba went to Binani Industries, a corporation representing a prominent Indian family with other non-mining interests in Zambia.[181] Binani optimistically promised no worker layoffs and significant investment in the mine's infrastructure.[182] Some of ZCCM's other assets then went to various companies that had formerly been part of the Kafue Consortium. Anglovaal bought what became Chambishi Metals while Phelps Dodge took over Zambia Metal Fabricators (ZAMEFA) in Luanshya. Anglo American once again assumed a more active role, purchasing the largest company, Konkola Copper Mines (KCM), based on hopes for Konkola Deeps. The mine privatization exercise then continued until 2000 when the sale of the Nkana and Mufulira sites to a small consortium running a company called Mopani Copper Mines completed the process. When all was finished, according to officials ZCCM's debts were actually more than cash receipts from the mines' sales covered. So, whatever revenue Zambia received at the time of sales went to pay off arrears.

The following year, Zambians voted Chiluba out of office. His attempts to remain in power by amending the constitution were the main focus of organized opposition against him. But weariness and unease over the privatization process may have figured in as well.[183]

A Re-privatized World

Despite the exhaustion and anxiety that accompanied the privatization process, eventual creation of new mining companies engendered an air of confident expectancy throughout much of Zambia. This optimism was, no doubt, enhanced by knowledge of Zimbabwe's simultaneous economic unraveling, for Robert Mugabe's farm invasion initiative had begun at roughly the same time that Nkana and Mufulira were sold. On the Copperbelt, however, trucks loaded with supplies and equipment regularly traversed the main roadways and small groups of miners walking to and from work became ubiquitous sights within the townships. Shift change sirens and periodic underground blasts kept even non-mining residents aware of the non-stop production cycle. The business produced by infusions of foreign cash was palpable and promising.

Yet, less than two years into the new ownership arrangement this outwardly robust environment revealed its inward fragility. Anglo American Corporation abruptly announced termination of major ownership in KCM, shocking the region and shaking the confidence even of the World Bank.[184] The reason, said company officials, was purely economic: copper prices had dropped so low and investment needs, particularly for KDMP, were so extensive that staying involved simply was not viable.[185] And, with Anglo's offices now in London rather than Johannesburg and Harry Oppenheimer's death the year before, the board was simply less inclined to stay.

Within a few months, a number of newer ancillary businesses such as restaurants and guesthouses began shutting down. Later the same year a second mine, RAMCOZ in Luanshya, also stopped production. Its owner, Binani, pulled out leaving, among other things, millions of US dollars' worth of unpaid terminal benefits due its workers. Miners staged demonstrations demanding settlement and at one dramatic public gathering even briefly held hostage the Province's deputy minister and the Minister of Mines.[186] It took over two years for new buyers to be found for both KCM and Luanshya.

By the time these were in place, the people had become even more suspicious and wary. The ultimate configuration of Zambia's remaining large parastatals, including Zambia Telecommunications Co. (ZAMTEL); Zambia Electric Supply Co. (ZESCO), and Zambia National Commercial Bank (ZANACO), became highly contested as passionate debate carried on in the popular press as to whether they should be left with the government, commercialized, or privatized.[187] But by mid-2005, 262 of 284 possible companies had been sold to private investors.[188]

In the aftermath of the entire process, the World Bank and IMF, architects of a process they had initially praised, acknowledged that miscalculations had been made along the way. Most of the lessons learned were, from the Bank's perspective, relevant to the realm of economics and management. More transparency and accountability should have been demanded, for instance. Clearer procedures should have been designed for personnel hiring and firing. But it also could not be expected, Bank officials mused, that privatization would provide Zambia anything more than a survival strategy. In the end, it might not prove the hoped-for catalyst of a comprehensive development strategy. Finally, deep within a few reports were scattered implications that the entire program really had not taken into account what Zambians thought about it all.[189]

What did they think? Many Zambians' expectations were that privatization would bring immediate benefit to the central bank and the country's economy. But, by the mid-2000s, with the first mine sales having taken place in the late 1990s, most people were still waiting for a promised better life. True, there were some business owners and managers who were pleased over their own increased business. But they also typically recognized that their vantage point was not shared by the average Zambian.

> With privatization they throw you all this jargon about how things are going to get better. We should look at the wider picture, but you will never see the wider picture if you're never totally at grips with what is on the ground. . . . Some people are doing much better. But what has it actually done for the majority of people who aren't head hunted? What has it done for them?[190]

Table 3.1. Major Change Periods in Zambia's Industrial Copper Mining History

1800s-1910s	**1920s-1930s**	**1970-72**	**1981-82**	**1997-2005**
Exploration	*Industrialization*	*Nationalization*	*Consolidation*	*Privatization*
British South Africa Company	Rhodesian Selection Trust *(USA & UK)* -Luanshya -Mufulira	Roan Consolidated Mines (RCM) -Mufulira -Luanshya -Baluba -Chibuluma -Chambishi	Zambia Consolidated Copper Mines (ZCCM)	Konkola Copper Mines[3] *(Vedanta Resources)* -Nchanga -KDMP -Chililabombwe
	Rhodesian Anglo American Corporation *(RSA)* -Rhokana -Nchanga -Bancroft	Nchanga Consolidated Copper Mines (NCCM) -Rokana/Mindola -Bwana Mkubwa -Nchanga -Konkola -Chililabombwe[1] -Kansanshi -Kabwe[2]		Mopani Copper Mines *(Glencore/First Quantum)* -Nkana[4]/Mindola -Mufulira
				Kansanshi[5] *(First Quantum)*
				Bwana Mkubwa *(First Quantum)*
				Chibuluma Mines *(Metorex)*
				NFC Africa Mining *(Nonferrous Metal Mining)* -Chambishi
				Luanshya Copper Mines[6] *(J and W Holdings)* -Baluba

[1] Previously known as Bancroft.
[2] Previously known as Broken Hill.
[3] At initial re-privatization, principally owned by Anglo American until 2002.
[4] Previously known as Rhokana and Rokana.
[5] Initially re-privatized by Cyprus Amax.
[6] Previously known as Roan Antelope and initially re-privatized by Binani Industries.

Figure 3.1. World Copper Prices 1927 to 2005

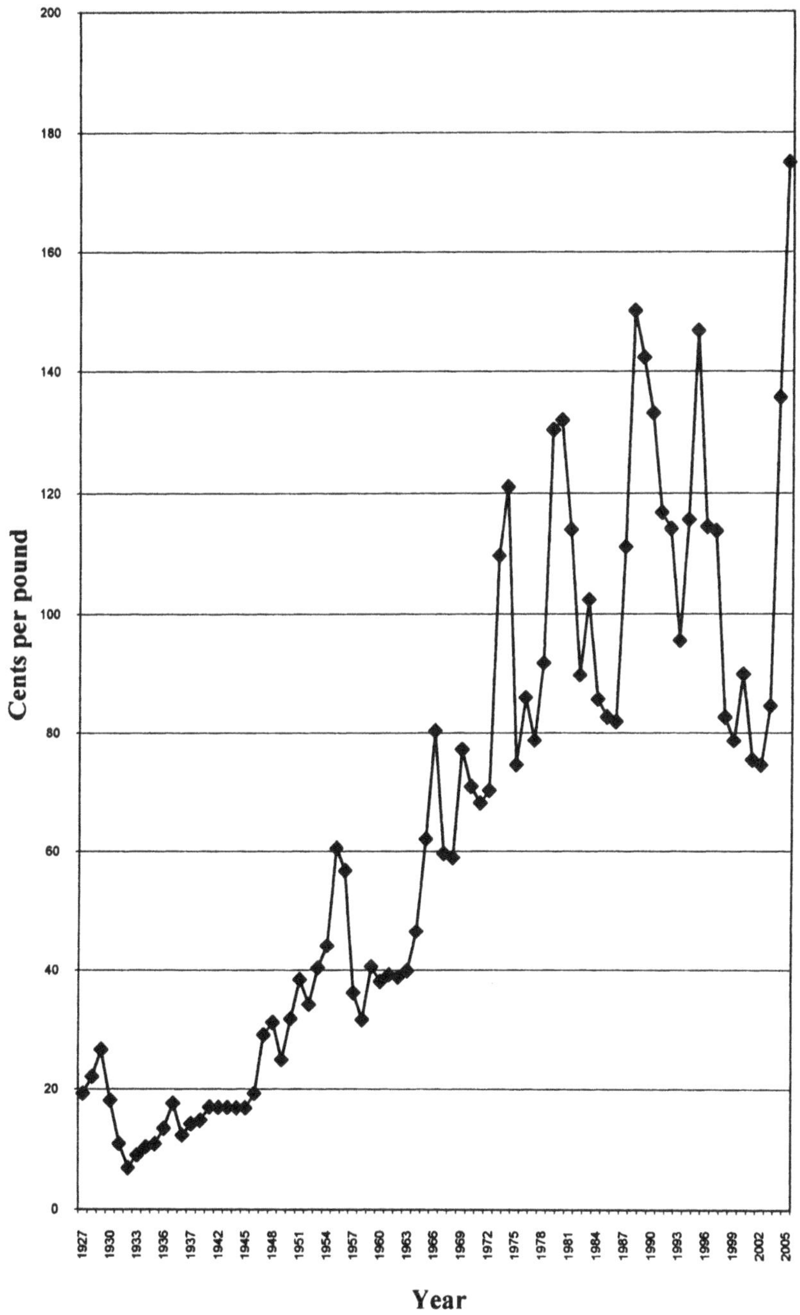

Source: Global Financial Database, courtesy Harvard University

Figure 3.2. Zambia Copper Production 1959 to 2005

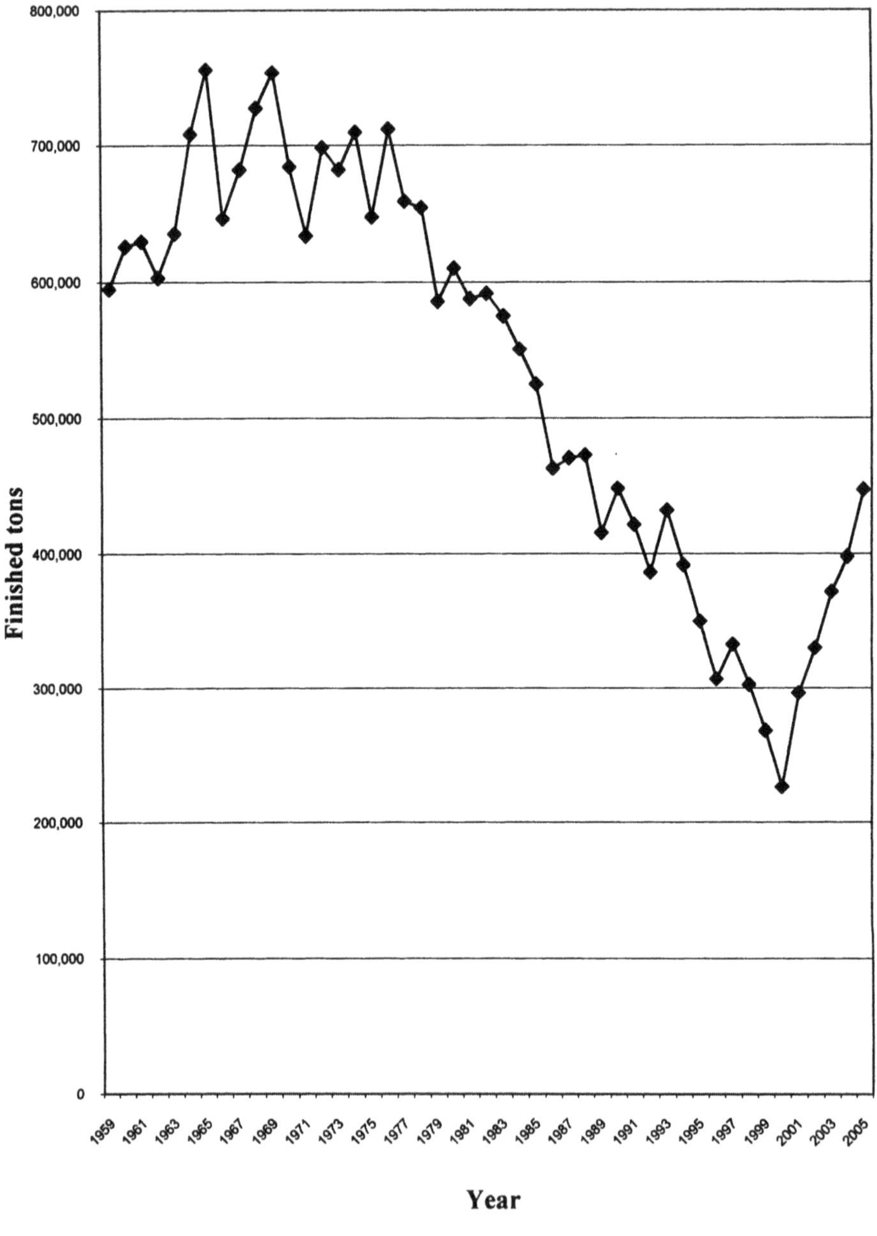

Source: Chamber of Mines, Kalulushi

Figure 3.3. Mine Personnel Statistics from ZCCM Consolidation to Privatization

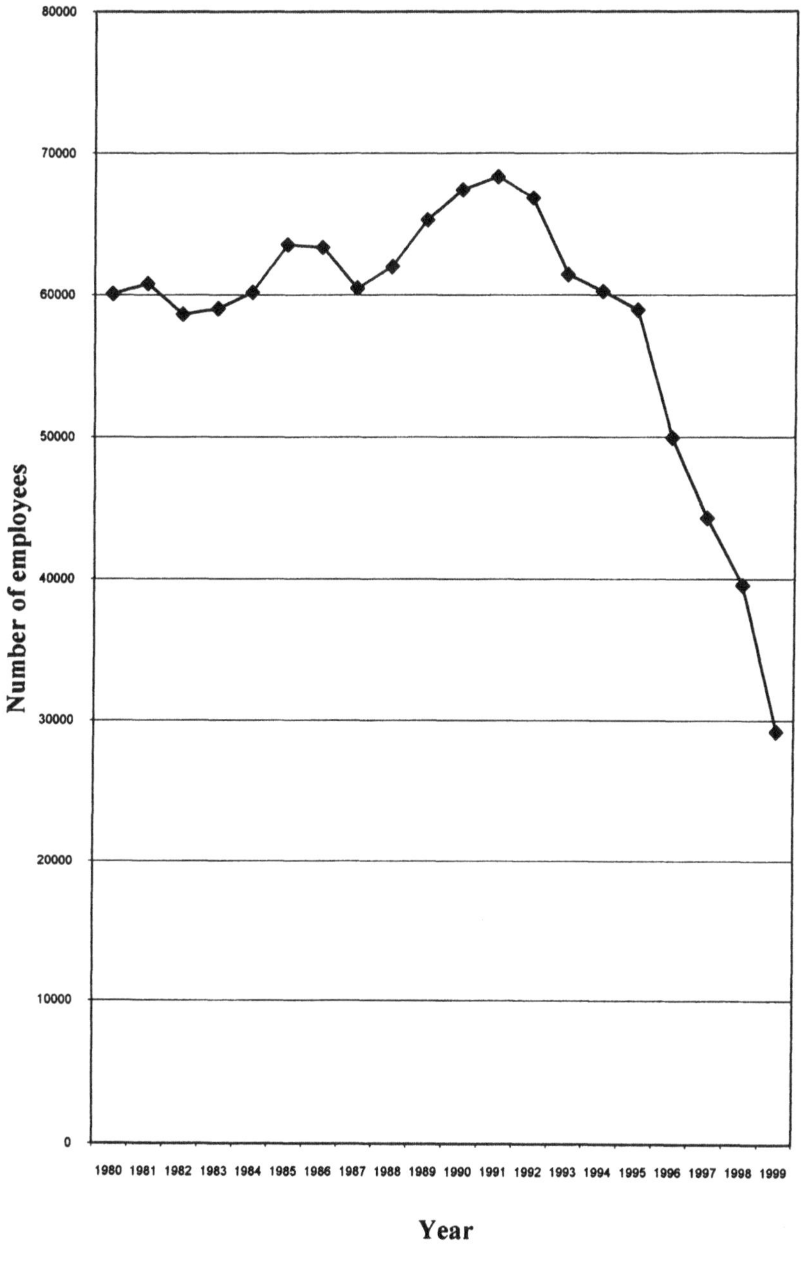

Sources: RCM, NCCM, ZCCM annual reports, Chamber of Mines, Kalulushi

Notes

1. Balcomb, "The Power of Narrative," 51.

2. The term "public" is used here to designate the sort of discourse that is presented and recorded through various media. It may be thought of similarly to the way in which public policy analysts talk about the "public square" or "public conversations." While, as has been argued from the outset here, there are always very important alternative viewpoints or narratives at work, a dominant, or pervasive (though dynamic), narrative is usually detectable in the ways that issues are defined, problems diagnosed, and so on. As Occhipinti's work has made clear, a "standard development discourse" can create a theoretical depiction of a country or cultural setting that is greatly at odds with lived experience. *Acting on Faith*, 47.

3. Roland Oliver and J. D. Fage, *A Short History of Africa* (London: Penguin Books, 1995), 19.

4. Oliver and Fage, *A Short History of Africa*, 109; Roberts, *A History of Zambia*, 86. See also Hall for a detailed description of the movements of different groups into this area during the pre-colonial era. *Zambia*, 5ff.

5. Radio carbon dating implies that "the whole complex process of mining, smelting, and working iron was independently invented in tropical Africa." In fact, some say the real reason why Europeans did not take advantage of the region's mineral resources sooner may have been because Africans were sufficiently well organized to run both the mining process and the metal trade. Oliver and Fage, *A Short History of Africa*, 1-2, 20, 111-12. Eugenia Herbert, *Red Gold of Africa: Copper in Precolonial History and Culture* (Madison: University of Wisconsin Press, 1984), 28, says that copper was widely used in Africa at least 1,500 years before colonialism. See also Cline, "The Sources of Metals," 54, 72; Felix Mendelssohn, ed., *The Geology of the Northern Rhodesian Copperbelt* (London: MacDonald, 1961), 3; Roberts, *A History of Zambia*, 36.

6. Cline, "The Sources of Metals," 46.

7. Cline, "The Sources of Metals," 53; Irving Kaplan, ed., *Zambia: A Country Study* (Washington, DC: The American University, 1979), 17. See also Hall for more traditional mining and trading history, *Zambia*, 245ff.

8. Oliver and Fage, *A Short History of Africa*, 110; Cline, "The Sources of Metals," 54.

9. Herbert, *Red Gold of Africa*, 150-53; *Zambia's Mining Industry: The First 50 Years* (Ndola: Roan Consolidated Mines, Ltd. 1978), 15.

10. Henry Hamilton as cited in Herbert, *Red Gold of Africa*, 154.

11. Stephen Neill, *The History of Christian Missions*, 2nd ed. (New York: Viking Penguin, 1986), 208-10.

12. Bonk, *Missions and Money*, 18.

13. Bosch, *Transforming Mission*, 227. The idea was that trade in human beings would lessen if trade in natural resources and goods could feasibly be undertaken.

14. Livingstone as cited in Kenneth Bradley, *Copper Venture: The Discovery and Development of Roan Antelope and Mufulira* (Great Britain: Mufulira Mines, Ltd. and Roan Antelope, Ltd., 1952), 31.

15. By 1890, Rhodes's chartered company had arranged its first mining concession with Lewanika, ruler of Barotseland and, by 1895, prospecting had begun north of the Zambezi River. The principal "Lewanika Concession" that the BSAC eventually used to claim ownership of all mineral rights was based upon corporate actions taken in 1900. See Mark Bostock and Charles Harvey, *Economic Independence and Zambian Copper: A Case Study of Foreign Investment* (New York: Praeger Publishers, 1972), xvi, 37. Roberts also notes that tribute and building of infrastructure such as schools and telegraph

lines promised by the BSAC to Lewanika in return for mineral rights were never delivered. *A History of Zambia*, 167. Additionally, Rhodes's plan for extending the British empire up from Southern Africa hinged on using a series of chartered companies similar to the BSAC from which capital could be drawn for development of the region. Powdermaker, *Copper Town*, 55-56. Another of Rhodes's major corporate endeavors, the De Beers Company, was founded as a consolidation of the South African diamond mines. De Beers and Anglo American Corporation, the mining house that figures so prominently in much of Zambia's history, have operated in tandem for decades. David Pallister, Sarah Stewart, and Ian Lepper, *South Africa, Inc.: The Oppenheimer Empire* (London: Simon and Schuster, 1987), 23.

16. Oliver and Fage, *A Short History of Africa*, 168; Roberts, *A History of Zambia*, 158, 162, 177.

17. Bostock and Harvey, *Economic Independence and Zambian Copper*, xvi; George Kay, *A Social Geography of Zambia: A Survey of Population Patterns in a Developing Country* (London: University of London Press, 1967), 15; Roberts, *A History of Zambia*, 161; Bradley, *Copper Venture*, 75. Amalgamation of the two territories came in 1911. *Zambia's Mining Industry*, 4. British interest in the area's resource potential heightened after World War 1 and that nation's desire for the ability to control its own copper supply. W. G. Watts, ed., "Commemorative Programme, 7th Commonwealth Mining and Metallurgical Congress, Northern Rhodesia Section" (Kitwe: Northern Rhodesia Section of the 7th Commonwealth Mining and Metallurgical Congress, 1961), 41. The protectorate arrangement was similar to that of Nyasaland's (present day Malawi) but considerably different from Southern Rhodesia's (present day Zimbabwe) arrangement as a self-governing colony of white settlers. Powdermaker, *Copper Town*, 59. The 1924 turnover was also the point at which what would eventually become the Anglo American Corporation first invested in the Copperbelt. By 1928, Rhodesian Anglo American (Rhoanglo) was incorporated in London. Pallister, Stewart and Lepper, *South Africa, Inc.*, 43.

18. It is also worth noting that African guidance in the search for copper deposits was not always willingly given. Bradley, *Copper Venture*, 32; Bostock and Harvey, *Economic Independence and Zambian Copper*, 53; Mendelssohn, *The Geology of the Northern Rhodesian Copperbelt*, 7. A pre-independence note to this effect is particularly rich in cultural assumptions. "The early prospectors were confronted with the difficulty that the natives were chary in imparting knowledge of minerals—a practice possibly forbidden by village headmen and the witch-doctors. It is thus not surprising that many of the first mineral discoveries were made almost by accident, though it should be recorded that early prospecting was assisted to an extent by native knowledge." Watts, "Commemorative Programme," 39.

19. Bostock and Harvey, *Economic Independence and Zambian Copper*, 26.

20. The Company's total deficit for the years 1891 through 1919 was £1.25 million. Buell, *The Native Problem in Africa*, 235. A federation-like concept that would help reduce the BSAC's administrative costs was considered from the early 1920s. At that time, however, the Company was virtually alone in its proposal for amalgamation since Europeans in Southern Rhodesia wanted to continue a process leading to self-rule while Europeans in Northern Rhodesia feared domination by their southern counterparts. Kaplan, *Zambia: A Country Study*, 25. Negotiations to turn administration of the territory over to Great Britain began in earnest in 1921 when European settlers rejected a Company plan to establish an income tax as a way of offsetting operating costs. According to some sources the Company, incurring deficits of approximately £130,000 annually, was quite willing to consider other arrangements for the territory and reached eventual agreement with Britain in 1923. Bradley, *Copper Venture*, 77.

21. Roberts, *A History of Zambia*, 192.

22. See also Taylor and Lehmann, *Christians of the Copperbelt*, 31.

23. Oliver and Fage, *A Short History of Africa*, 167. Kay characterizes the mining communities of that era as consisting "largely of hard-headed Europeans and bewildered Africans who . . . succumbed easily and in large numbers to disease." *A Social Geography of Zambia*, 136.

24. "Copper and the Copper Mining Industry of This Country" (photocopy, ZCCM Archives, Ndola, 1964), 12-13; *Zambia's Mining Industry*, 25-29; Kaplan, *Zambia*, 187. Taylor and Lehmann observe that RST had "considerable American influence" and further contend that the AAC's title was "misleading" given the overwhelming influence of South African interests. *Christians of the Copperbelt*, 29.

25. See White for historical details on how the Great Depression affected the lives of Copperbelt workers. *Speaking with Vampires*, 270.

26. *Zambia's Mining Industry*, 34; Mendelssohn, *The Geology of the Northern Rhodesian Copperbelt*, 9; "Copper and the Copper Mining Industry," 12.

27. James Eric Lane, *Moment of Encounter* (New York: Peter Lang, 1984), 13.

28. Lane, *Moment of Encounter*, 51; Helmuth Heisler, *Urbanisation and the Government of Migration: The Interrelation of Urban and Rural Life in Zambia* (London: C. Hurst and Co., 1974), 4-5, 37.

29. Bostock and Harvey, *Economic Independence and Zambian Copper*, 27; Roberts, *A History of Zambia*, 177-78. The tax was about 10 shillings per year. Payment for farm work could run about 4 pence and on the mines 6 pence. Roberts, *A History of Zambia*, 24. Taylor and Lehmann comment on the double-edged nature of this development. On the one hand, the hut tax did help precipitate rural impoverishment and the breakdown of family discipline. On the other, those who traveled to the Copperbelt for economic reasons gained experiences that eventually led to an emergent African leadership. They liken this situation to a "tension between the mystical appeal of the tribal head and the new powers of the 'young educated,'" which, they contend, had a huge historical and cultural impact on the Copperbelt. *Christians of the Copperbelt*, 33.

30. Lane, *Moment of Encounter*, 13.

31. Roberts, *A History of Zambia*, 178. See also Bostock and Harvey, *Economic Independence and Zambian Copper*; Heisler, *Urbanisation and the Government of Migration*; Ferguson, *Expectations of Modernity*; and Kay, *A Social Geography of Zambia*.

32. See Taylor and Lehmann for an extensive discourse on this period. *Christians of the Copperbelt*, 33-34.

33. Officially founded in 1958, the center later changed its name to Mindolo Ecumenical Foundation. While the grounds are not far from the Mindola mine shaft, Mindolo was known as an easily accessible shady spot outside Kitwe where African representatives from congregations around the Copperbelt would assemble for ecumenical meetings. See Taylor and Lehmann, *Christians of the Copperbelt*, for an extensive history of this era, 37-50. See also Denis M'Passou, *Mindolo: A Story of the Ecumenical Movement in Africa* (Lusaka: Multimedia Press, 1983), 3; and Gladys Jamieson, *Zambia Contrasts: A Page in the Story of Christian Education in Zambia* (London: London Missionary Society, 1965). Jane Parpart offers a somewhat different perspective on these developments. She sees the collaborative efforts to provide training and resources for mining families as flowing from the companies' desires for increased control over the workforce following the strikes of 1935 and 1940. "Prodded by colonial office and Northern Rhodesian officials, management at both Copperbelt companies decided to support programs to teach women useful skills which could stretch their husbands' meager wages while keeping them busy and out of mischief. The mines turned to the United Missions in the Copper-

belt (UMBC), an ecumenical urban-based mission group concerned with the education and welfare of Africans on the Copperbelt." Jane L. Parpart, "Class and Gender on the Copperbelt: Women in Northern Rhodesian Copper Mining Areas 1926-1964" (Working Paper No. 77, African Studies Center, Boston University, 1983), 8. Regardless of the mix of motives behind any of the colonial-based institutions—from corporations to churches—the point being made here and in Parpart is that African mining families were still able to turn elements of the situation to their own advantage. That is one of the remarkable features of their resilience under oppressive circumstances.

34. According to Taylor and Lehmann, there was a "sudden imposition of increased taxation." *Christians of the Copperbelt*, 39. See also Epstein, *Politics in an Urban African Community*, 29. Meebelo, however, asserts that the tax increase may have simply been the spark that set alight a smoldering fire of resentment. In his thorough treatment of the 1935 riots, he documents a list of grievances African workers had compiled that included under payment and beatings. See *Reaction to Colonialism*, 254. See also the study of labor and living conditions on the Copperbelt, as well as African miners' strategies for creating ingenious coping strategies in Albert B. K. Matongo, "Popular Culture in a Colonial Society: Another Look at Mbeni and Kalela Dances on the Copperbelt, 1930-64," in *Guardians in Their Time: Experiences of Zambians Under Colonial Rule, 1890-1964*, ed. Samuel N. Chipungu (London: Macmillan Press, 1992), 180-217.

35. Robert Rotberg as cited in Meebelo, *Reaction to Colonialism*, 254. White contends that the strikes "had a profound impact on British imperialism" in that they prompted unprecedented anthropological investigations into mining labor practices and urban life. *Speaking with Vampires*, 272.

36. The Watch Tower movement was an early 20th century off shoot of what eventually became the Jehovah's Witnesses. A prophetic and charismatic movement stressing a millennial second coming of Christ, it was eventually viewed as threatening to the British colonial powers. David Chidester, *Christianity: A Global History* (San Francisco: HarperSanFrancisco, 2000), 420-24. A Watch Tower revival in 1907 in Nyasaland was characterized by prophecies that the world would soon end, thus bringing an end to colonial rule and the beginning of an era of peace and harmony. Fields, *Revival and Rebellion*, 5. In his investigation into pre-independence political activity in Northern Zambia, Meebelo discusses the case of six Africans who had been deported from Southern Rhodesia allegedly because of their involvement with Watch Tower. The noteworthiness of the deported six Africans, though, is due to the ways in which their views and activities were shaped by experiences of economic inequity in Southern Rhodesia. At least in the Northern regions, Watch Tower was, says Meebelo, "an emotional safety valve through which the African people vented their frustrations under white rule, and solaced themselves with hopes that some day Providence would, through the leaders of the movement, put things right." See his exhaustive account of this in *Reaction to Colonialism*, 133-85. See also the account of these events in Fields, *Revival and Rebellion*, 138-43.

37. "Report of the Commission Appointed to Enquire into the Disturbances in the Copper belt, Northern Rhodesia Together with the Governor's Despatch to the Secretary of State on the Report, Herbert Young, Governer" (n.p., November, 1935), 4.

38. Meebelo, *Reaction to Colonialism*, 174. At the time of the 1935 disturbances, there were 855 Europeans and 6,556 Africans at Nkana; 383 Europeans and 3,078 Africans at Mufulira; and 595 Europeans and 4, 442 Africans at Luanshaya—the three areas involved in work stoppages and demonstrations. "Report of the Commission Appointed to Enquire into the Disturbances," 10, 17, 20. Evidence that such disproportionate representation played a part in European nervousness about their own situation derives from the force with which the "disturbances" were met. While work stoppages seemed to con-

stitute the extent of the display in Mufulira and Nkana, at Luanshya about 1,000 Wemba (i.e., Bemba) "natives" were said to have attacked the mine compound. The Europeans opened fire, killing six and wounding others. The firing of arms came after a "native" reportedly struck an armed soldier. Additionally, the Royal Air Force was called up from Salisbury along with 54 European police from Salisbury; 25 European police; and 35 "native" police from Bulawayo. There was also talk of securing an airplane that could carry tear gas bombs and the police patrolled the compounds for some time thereafter. "Report of the Commission," 7-9. The secretive nature of Watch Tower was, according to Fields, a primary reason why colonial officials remained concerned about it since adherents might be able to hide conspiracies against whites. *Revival and Rebellion*, 192.

39. In 1949, the various African unions came together to form Northern Rhodesia African Mineworkers' Union directed by Lawrence Katilungu. Kaplan, *Zambia*, 31. European mine workers had their own representative bodies. Geoffrey W. Silavwe, *Some Aspects of Personnel Management Practice in the Copper Mining Industry of Zambia* (Ndola: Mission Press, 1999), 298.

40. For the decade 1930-40 approximately £2,400,000 worth of taxes flowed out and only £136,000 returned in development funds. During the colonial era, the mines paid taxes to Salisbury in Southern Rhodesia and a portion of that tax was returned to Northern Rhodesia for province administration. Roberts, *A History of Zambia*, 193.

41. Roberts, *A History of Zambia*.

42. Roberts, *A History of Zambia*. The United Kingdom was in debt to the United States and sought high copper production from its colonies to help offset wartime expenses. "Copper and the Copper Mining Industry," 10; Watts, "Commemorative Programme," 41.

43. The 1951 census recorded almost 20,000 Europeans and close to 90,000 Africans as mine employees. Europeans were immigrating to the area at a rate of almost 8,000 per year. Correspondingly, the Northern Rhodesian government's income rose from £856,376 in 1931 to £15,632,472 some 20 years later. Production figures went from 16,691 long tons in 1931 to 309,142 in 1951. Bradley, *Copper Venture*, 24-25.

44. Silavwe, *Some Aspects of Personnel Management*, 5; Ferguson, *Expectations of Modernity*, 49; Heisler, *Urbanisation and the Government of Migration*, 2; Gavin W. H. Relly, "Background to the Zambian Copper Mining Industry," An Address by the Chairman of Anglo American Corporation (Central Africa), Ltd. On the Producers Day at the Inter-governmental Copper Conference (Lusaka, 1967), 6.

45. Bradley, *Copper Venture*, 22-23. Outside consultants with worldwide mining experience who came to the Copperbelt in the 1980s remarked that they had never seen anything comparable elsewhere. See also Matongo, "Popular Culture in a Colonial Society," 184-91.

46. Heisler, *Urbanisation and the Government of Migration*, 114. White's historical research also reveals that African miners began migrating from job to job and/or mine to mine rather than simply taking long rural sojourns between job assignments. *Speaking with Vampires*, 288.

47. Anglo American annual report for 1951 as cited in Pallister, Stewart, and Lepper, *South Africa, Inc.*, 42. See Taylor and Lehmann for details of the town-mine relationship. Generally, the town or municipality was the area in which non-mine employees resided. The "mine" actually referred to everything that the companies owned including underground development, recreation facilities, schools, hospitals, and houses. *Christians of the Copperbelt*, 30.

48. Taylor and Lehmann report, however, that an apartheid type system did exist during the time of their research. For example, the mine and municipal areas were bi-

sected into European and African areas and members of one race were "forbidden" to live in areas allocated to the other. By the time of their writing, there were no restrictions on who could enter which shop but such restrictions had been in place in the past. *Christians of the Copperbelt*, 30.

49. Kay, *A Social Geography of Zambia*, 136. Taylor and Lehmann refer to this as "the omni-competent paternalism of modern industry." *Christians of the Copperbelt*, 30.

50. As quoted in Pallister, Stewart, and Lepper, *South Africa, Inc.*, 45-46. By the time Jamieson was writing about the Copperbelt (early 1960s), she reported that in Luanshya "three-, five-, and even seven-roomed houses were being built to replace the old rows and rows of one-roomed huts. There were gardens and open spaces for the children. We heard of highly-paid Africans whose status entitled them to live in erstwhile 'Europeans-only' areas." *Zambia Contrasts*, 70.

51. In addition to Watch Tower, Alice Lenshina Mulenga and the Lumpa Church are discussed at length in Taylor and Lehmann, *Christians of the Copperbelt*, 248ff. See also the extended analysis of Watch Tower revivals along with the colonizers' interpretations of and responses to them that took place throughout Northern Rhodesia from the early 1900s onward in Fields, *Revival and Rebellion*.

52. An extended account of this crucial period in Zambian history may be found in Taylor and Lehmann, *Christians of the Copperbelt*, 121ff.

53. Kaplan, *Zambia*, 32; Anthony St. John Wood, *Northern Rhodesia: The Human Background* (London: Pall Mall Press, 1961), 134.

54. Noting the "tragedy" of Africans' propensity to distrust even the goodwill of Europeans, Taylor and Lehmann go on to observe that, "the pattern of society—which this well-meaning majority [of Northern Rhodesians] too readily accepts—has not made it easy for a member of one race ever to discover what the members of another are really thinking." *Christians of the Copperbelt*, 128.

55. Epstein saw Federation as not only supported by the BSAC but directly related to the Company's handover to the British government some 31 years previously, *Politics in an Urban African Community*, 158. Anglo American and other mining companies with British and American interests were all enthusiastic supporters of Federation. Pallister, Stewart, and Lepper, *South Africa, Inc.*, 44. Frederick Cooper asserts that Great Britain hoped Federation would halt the rising power of South Africa's white ruling class from taking control of more territory within the region. Frederick Cooper, *Africa Since 1940: The Past of the Present* (Cambridge, UK: Cambridge University Press, 2002), 74. See also Wood's discussion of the issue in *Northern Rhodesia*, 68.

56. Taylor and Lehmann, *Christians of the Copperbelt*, 121. Federation was vehemently opposed by the majority of Africans. Anthropologists and historians who have studied this period closely use quite strong language to describe what they observed. "Fear of losing land was widely shared among Africans in Northern Rhodesia. Of all their reasons for opposing amalgamation or federation, this fear carried the greatest emotional force, and it crystallized other fears which were perhaps less easily defined." Roberts, *A History of Zambia*, 209. "When I began my field work, the meaning of the new political union became quickly apparent to me because the fear of federation and the loss of land was dragged into almost every interview and conversation, regardless of context or relevancy." Powdermaker, *Copper Town*, 63-64. "[T]he Federation issue produced on Africans an impact akin to trauma." Epstein, *Scenes from African Urban Life*, 172-77. White argues that the depth of feeling and distrust of Europeans may be reflected in rumors that circulated at the time. These included stories regarding Europeans poisoning sugar and preparing tinned meat made with human flesh, all for consumption by Africans. *Speaking with Vampires*, 297-99. See also Powdermaker, *Copper Town*, 644.

57. Taylor and Lehmann, *Christians of the Copperbelt*, 126-30.

58. From 1954/55 to 1959/60, the net drain on Northern Rhodesia due to taxes on the mines for Federation was £56 million. At the same time, Northern Rhodesia could only spend £50 million on capital expenditures within its own territory and, in all, lost £97 million in taxes. Martin, *Minding Their Own Business*, 36.

59. A book and even documentary film were made of this endeavor. Charles Lagus, *Operation Noah* (London: William Kimber and Co., Ltd., 1959). A few other historical points of note concerning Kariba Dam deserve mention. The £20 million cost of construction was, in part, funded by the Anglo group of mines. Pallister, Stewart, and Lepper, *South Africa, Inc.*, 45. In the 1970s as tensions between Rhodesia and Zambia increased, the World Bank funded construction of a power station on the north side of the Zambezi that was understood as an effort to give Zambia more autonomy over its electrical power. Peter Stiff, *See You in November* (Alberton: Galago Publishing, Inc., 1985), 85-86. In the mid-2000s, this power station was owned by one of Zambia's remaining parastatals, Zambia Electric Service Co., or ZESCO. Further, the population displacement ripple effect was still apparent in small, but important details. One concerned the manner in which Tonga people were able to cross the Zimbabwe-Zambia border. They could move between the two countries using just certain paper documentation rather than having to obtain official passports. Another interesting detail concerned responsibility for maintenance and upkeep of the Dam itself. This had not been such an issue under Federation or even for decades afterward as long as Zimbabwe was flourishing. But by the mid-2000s, with that country's economic and political stability shattered, questions began to arise concerning which country actually owned the Dam.

60. As Roberts observes, "Fear of losing land was widely shared among Africans in Northern Rhodesia. Of all their reasons for opposing amalgamation or federation, this fear carried the greatest emotional force, and it crystallized other fears which were perhaps less easily defined. Land was still of absorbing interest to almost all Africans. The growth of industry on the Copperbelt had done nothing to undermine this interest; rather, indeed, it had strengthened it." *A History of Zambia*, 209.

61. European farming had rapidly expanded in Southern Rhodesia after World War I and tens of thousands of Africans from Northern Rhodesia had migrated south over the years for mine work elsewhere. Roberts, *A History of Zambia*, 91.

62. By 1960, only about half of all white expatriates were staying in Northern Rhodesia for the duration of their work lives; and only about 16% of the expatriate population retired on the Copperbelt. This was largely due to the cost of living and being in an undeveloped area with not a lot of amenities for seniors. National Institute for Personnel Research, in Association with the Institute for Social Research, *The Attitudes of White Mining Employees Towards Life and Work on the Copperbelt* (Pietermaritzburg: University of Natal, 1960), i. Adding to the impermanence of the white population there was the fact that the Northern Rhodesian mines had to compete with those in South Africa. This meant offering greatly subsidized wages and amenities that produced an artificially enhanced standard of living for Europeans who came to the Copperbelt. Taylor and Lehmann, *Christians of the Copperbelt*, 32.

63. At least one contemporary observer credited Federation with deepening racism between Africans and Europeans because, prior to amalgamation, inter-racial relationships had been looked upon primarily as being between employer and employee more so than colonial settler and indigenous African. Taylor and Lehmann, *Christians of the Copperbelt*, 137.

64. Kaplan, *Zambia*, 33; Powdermaker, *Copper Town*, 62.

65. UNIP was led by Kenneth Kaunda who was to become Zambia's first president. But Jamieson, in describing "the troubles," counters a government white paper of 1961 that attributed sporadic violence to his political party. She credits him personally with projecting a consistent message of non-violence. *Zambia Contrasts*, 73.

66. Roberts, *A History of Zambia*, 190.

67. Powdermaker reports that she encountered a number of European women who were genuinely interested in learning more about Africans but were inhibited from doing so by systemic race and class divisions. *Copper Town*, 69.

68. The sparse European presence on the ground during the period of Indirect Rule, which began officially for Northern Rhodesia in 1929, was one of the policy's most remarkable traits. By 1960, when the British administrative presence was at its height in Zambia, the United Kingdom's government was represented by 274 men in a territory of 290,000 square miles. Fields, *Revival and Rebellion*, 33-34. Because the colonial governments had limited interests in their territories, they deliberately kept representation there small. Appiah, *In My Father's House*, 164.

69.Taylor and Lehmann also suggest that the problem of governance and management of infrastructure and resources had unfortunate cultural and religious roots. "Since the old African ways, in which religion and society are one, met our schizophrenic Western culture (built paradoxically, as it is, upon a materialistic technology and the Christian faith), many Africans, failing to learn the trick of holding the religious and the secular spirit in separate compartments, have abandoned themselves to a meaningless immediacy—"eat, drink and be merry, for tomorrow we die." *Christians of the Copperbelt*, 275.

70. Martin, *Minding Their Own Business*, 35; Roberts, *A History of Zambia*, 218. African girls particularly suffered from lack of educational opportunities. Davis, *Modern Industry and the African*, 388.

71. It bears further note that, within an African population of approximately three million and of the few schools that did accommodate secondary education, only one was a university preparatory school. Pallister, Stewart, and Lepper, *South Africa, Inc.*, 45. Mission schools helped address some of this gap. Silavwe, *Some Aspects of Personnel Management*, 2; Adu Boahen, *African Perspectives on Colonialism* (Baltimore: Johns Hopkins University Press, 1987), 86. Jamieson comments, "Education in Northern Rhodesia was something to be struggled after." By her calculations, in 1963 only one in approximately seventy children who started school was able to get to the secondary level. *Zambia Contrasts*, 45.

72. "Roan Consolidated Mines Limited" (photocopy, ZCCM Archives, 1980), 9; Kapya John Kaoma, "The Democratic Crisis in Southern Africa: A Challenge to the Church" (unpublished book manuscript, Mutare, Zimbabwe, 2001), 12. Roberts, citing these same figures, notes that Uganda achieved similar educational benchmarks in 1955 and Ghana in 1943. *A History of Zambia*, 234. According to Jamieson, however, recorded demographic numbers could be suspect because of the fluidity of African life. Many births and deaths, for example, took place beyond the knowledge of any colonial institution. *Zambia Contrasts*, 47. "It is the most damaging criticism of all against the colonial and federal governments that Northern Rhodesia, despite its possession of the richest economy in Africa north of the Limpopo, found itself at independence with a smaller number of educated Africans in relation to the population than virtually any other of Britain's African colonies." Martin, *Minding Their Own Business*, 49.

73. Only a small percentage of Zambians were eligible to vote in the 1962 elections and many of those who were expressed frustration with the complicated voting system. Each elector, for example, had to vote for three things—an upper roll or a lower roll candidate, and two votes for national seats. Jamieson, *Zambia Contrasts*, 76.

74. Robert Bates, *Patterns of Uneven Development: Causes and Consequences in Zambia*, Monograph Series in World Affairs 11, no. 3 (Denver: University of Denver, 1974), 19.

75. In 1950 the BSAC had negotiated to keep all mineral royalties until 1986. Bostock and Harvey, *Economic Independence and Zambian Copper*, 43. The year prior to Zambian independence, mineral royalties earned the Company a net £6 million, twice the budget that went into African education. Roberts, *A History of Zambia*, 222.

76. Roberts, *A History of Zambia*, 48; "Roan Consolidated Mines," 108-09. The following year, Charter Consolidated was created by a merger of the BSAC, Central Mining, and Anglo American's original parent company, Consolidated Mines Selection. The BSAC brought to this merger the money from its 11th hour settlement back in Lusaka. Charter Consolidated went on to become a holding company for operations in the USA, Britain, Canada, and France. Pallister, Stewart, and Lepper, *South Africa, Inc.*, 82-83.

77. Kaunda's parents were Nyasalanders. His father, David, had come to Northern Rhodesia from the Livingstonia mission and eventually became well known as a Scots Presbyterian minister and educator at Chinsali's Lubwa Mission. Earlier in the industrial mining era, the companies had begun recruiting Africans from Nyasaland who had had longer exposure to Europeans than had Africans living in Northern Rhodesia. African miners from Nyasaland also were often products of mission schools and this combination of experiences meant that they fairly quickly assumed responsible positions such as training and clerical jobs within the mines. Workplace leadership translated into church leadership as well and the impact of Nyasaland Africans upon the development of Zambian church and society was significant. Taylor and Lehmann, *Christians of the Copperbelt*, 33. See also Hall, *Zambia*, 51. In addition to his own formal training in education, Kenneth Kaunda worked at a couple of Southern Rhodesian mines. The lack of the people's preparedness for governance was something he seemed never to forget. "[O]ur rulers: seventy years they produced one hundred university graduates; one hundred university graduates," he recalled decades later. Interview with author, 2005. The views he expressed over the way education and business had been set up by colonial interests were similar to those expressed by other Africans from around the same time. See, for example, Fanon, *The Wretched of the Earth*, 190 and Aimé Césaire, "Discourse on Colonialism," in *African Philosophy: An Anthology*, ed. E. Chukwudi Eze (Malden, MA: Blackwell Publishers, 1998), 222-27.

78. See footnote 65. Perceptions of the level of violence created in Northern Rhodesia varied among Westerners who recalled that period and who were interviewed for this study. Some contended that the transition to independence had come peacefully, disturbed at most largely by rock throwing. Others attributed coordinated, periodic violent episodes to UNIP Youth and still others recalled a violent trend lingering on into independence that displayed itself in acts of ecological irresponsibility. These included widespread deforestation as part of a government led charcoal producing initiative and mass slaughter of game animals for entertainment and dining at State occasions.

79. Stiff, *See You in November*, 101; Kaoma, "Democratic Crisis in Southern Africa," 13-14.

80. Pallister, Stewart, and Lepper, *South Africa, Inc.*, 178-79.

81. Martin, *Minding Their Own Business*, 115. Indeed, early indications were that the nationalist government and private copper mining companies would work well together. Hall, *Zambia*, 264.

82. One technical consultant who worked on the ZCCM privatization project observed that the mines under government ownership never profited sufficiently to warrant dividend distribution to Anglo. During the privatization process, the Zambian govern-

ment was averse to the idea of Anglo's re-engaging beyond its minority shareholder position.

83. For example, Kaunda took a prominent role in HIV/AIDS prevention and care.

84. Ellis and Ter Haar, *Worlds of Power*, 70.

85. Kaunda early on defined Humanism this way. "I have a passionate belief in the worth and possibilities of man and I expect him some day to achieve perfection. By perfection I do not mean sinlessness. But for all his weaknesses, man is growing in self-knowledge and will one day fully realize his capabilities." Kenneth Kaunda, *A Humanist in Africa*, 19. See also Kenneth Kaunda, "The Imperative of Human Dignity," Address to the Inaugural Session of the Non-Governmental Organisations International Conference on Human Rights, Paris September 15, 1968 (Lusaka: Zambia Information Service, 1968), 3. Elsewhere, he delineates the three key factors of tribal society as 1) mutuality, 2) acceptance, and 3) inclusiveness. *A Humanist in Africa*, 24-26.

86. For more on the Cold War's effects on sub-Saharan Africa see Peter Schwab, *Africa: A Continent Self-Destructs* (New York: Palgrave Macmillan, 2001). Martin's *Minding Their Own Business* also contains an excellent recounting of the background for Zambia's transition to a socialist state. According to his study, a combination of factors and events including Zambians' general lack of education and skills training and Kaunda's own disillusionment with Western government and business interactions after 1964 eventually helped "turn Kaunda from an optimistic economic liberal to an increasingly tough interventionist." To a certain degree, the decision to nationalize the mines arose, then, more from political and emotional rather than purely economic motivations. *Minding Their Own Business*, 95, 199. Paradoxically, writing in the same year, Burawoy maintains that the Kaunda government's desire not to interfere with the mines' profitability led it to neglect instituting a comprehensive Zambianization program. Instead, the government demonstrated lack of sympathy for mineworkers and their unions. *The Colour of Class*, 104.

87. Some Zambians assert that the entire African independence movement should be interpreted from a socialist point of view in that political independence was an opportunity to reclaim the means of production from those who had been illegitimately taking the land and the people's wealth. In Zambia's case, for example, during the ten years immediately preceding independence, the two major mine company owners in Northern Rhodesia had sent £260 million out of the country in dividends, interest and royalty payments. From 1923 to 1964, the BSAC received £82 million net for its mineral royalties. Pallister, Stewart, and Lepper, *South Africa, Inc.*, 84.

88. Kaplan, *Area Handbook for Zambia*, 202.

89. Kaoma, "The Democratic Crisis in Southern Africa," 18. There was also very little public discussion about copper and the mines in the early days of Zambia's existence. This was in marked contrast to the situation in other countries such as Chile. Martin, *Minding Their Own Business*, 132. Martin also states that Humanism's weakness was its vagueness and lack of applicability to concrete situations. *Minding Their Own Business*, 107.

90. Martin, *Minding Their Own Business*.

91. "Kaunda, like most national leaders in the third world had a generalized view of how things should be in this new independent state. In the quest of addressing the issue of under development, he developed an ideological program, which combined economics, politics, social relations, spiritual and international relations with the hope of transforming the country within the shortest possible time. Unfortunately, this vision does not fit in the grand vision of the State and though it blindly convinces the masses that a major transformation is possible, experience shows that such programmes are short lived."

Kaoma also asserts that this opposition took place in part because of the Churches' own foundations in Western Christianity. "The Democratic Crisis in Southern Africa," 17, 21. See also Zambia Episcopal Conference, Christian Council of Zambia, and Zambia Evangelical Fellowship, *Marxism, Humanism, and Christianity* (Lusaka: Teresianum Press, 1979).

92. President Kenneth Kaunda, interview with author, 2005.

93. See the discussion in chapter one as well as Deneulin, *Religion in Development*, especially chapter three.

94. One of Deneulin's main points is that the "public face" of religion in developing countries "can neither be ignored nor contained within certain boundaries." *Religion in Development*, 6.

95. UDI was declared in November, 1965. Similar to Northern Rhodesia, Nyasaland had become independent Malawi in 1964.

96. Ultimately, Zambia helped free neighboring countries that then began their independence eras in better financial shape than Zambia found itself. Kaoma, "The Democratic Crisis in Southern Africa," 15. The country's military budget increased four fold from the late 1960s through early 1970s. Armed forces grew from 5,000 in 1964 to 16,000 by 1973. Military expenditures as a percent of GNP doubled by the early 1970s and arms imports were more than $50 million USD by 1973. Kaplan, *Zambia*, 43. During the early days, the effects of this situation were not so apparent on the white community that remained following independence. In fact, some who participated in this study recalled that era with a bit of nostalgia, for curfews imposed to accommodate African rebel troop movements became invitations to weekend house parties. As long as everyone bought enough supplies in advance and cleared off the roads before dark, an official curfew had little residual effects on them.

97. Anglo's transportation company, Freight Services, also supported Ian Smith's regime by helping oil companies evade British sanctions under UDI. Pallister, Stewart, and Lepper, *South Africa, Inc.*, 66. See also Stiff, *See You in November*.

98. Hall points out that geography also contributed to the difficulties of Zambia's infrastructure development and its subsequent hardships during the UDI period. Most African countries had been approached by colonialist development from the sea inward. But building communication and transportation mechanisms for landlocked Northern Rhodesia was particularly costly. *Zambia*, 268.

99. This agreement with China was the first major arrangement of its kind between the two countries and further entrenched the Kaunda government's relationship with Communist powers. But, according to technical experts familiar with this history, some Western sources had said the pipe was impossible to construct.

100. Martin, *Minding Their Own Business*, 52-53.

101. "Roan Consolidated Mines," 4; *Annual Report of the Mines Department for the Year 1968* (Lusaka: Government Printer, 1969), 1.

102. This incident and the reasons behind it are carefully documented in Martin, *Minding Their Own Business*, 79ff.

103. Bostock and Harvey, *Economic Independence and Zambian Copper*, 100.

104. In 1968, President Kaunda stated that the mining companies had been distributing over 80% of their annual profits as dividends and had not been recapitalizing the mines. Bostock and Harvey, *Economic Independence and Zambian Copper*, 123.

105. Alastair Heron, *Accidents in the Zambian Mining Industry* (Lusaka: Government Printer, 1968), 17.

106. Roberts, *A History of Zambia*, 229; Bostock and Harvey, *Economic Independence and Zambian Copper*, 91; "Roan Consolidated Mines," 5.

107. Roberts, *A History of Zambia*, 231. Economic diversification was also an area in which the Kaunda government was at a disadvantage due to colonial era policies. Throughout the period of BSAC administration and British protectorate status, the territory's development had been largely confined to the rail line leading from Southern Rhodesia up to Katanga. This meant that the route from Livingstone, through Lusaka and Kabwe and into the Copperbelt was most favored for infrastructure improvements while the vast expanses where rural dwellers stayed was largely ignored. Parastatal organizations instituted by the new Zambian government pressed President Kaunda for more and more subsidization as there were few means for establishing profitable enterprises in the rural areas. See Bates, *Patterns of Uneven Development*, 15ff, for a comprehensive discussion of problems inherent in Zambia's early urban-rural development dilemma.

108. They were subsidiaries of Rhodesian companies. Ravi Gulhati, "Impasse in Zambia: The Economics and Politics of Reform" (EDI Development Policy Case Series, Analytical Case Studies no. 2, The World Bank, Washington, DC, 1989), 14.

109. Kaplan, *Zambia*, 188. Martin, *Minding Their Own Business*, 94-97.

110. Kaunda may have been prepared for direct and public opposition because he had disbanded the National Council several days earlier and instituted security measures immediately following the Matero speech. Martin, *Minding Their Own Business*, 156-57. Government pamphlets subsequently assured the public that the strike bans and wage freezes were temporary. *Towards Complete Independence: What it all Means* (Lusaka: Government Printer, 1969), 7. There was also in 1969 a clash between UNIP and members of the Watch Tower movement in Luapula Province. Watch Tower by that time was being seen in some quarters as "an instrument of foreign subversion against Zambia." Kaunda subsequently banned Watch Tower from the country. Meebelo, *Reaction to Colonialism*, 173.

111. *Towards Complete Independence*, 4-5. Writing the year before and from an entirely different country Fanon nevertheless analyzed nationalization quite prophetically. He observed that to ordinary Africans nationalization would represent, "the transfer into native hands of those unfair advantages which are a legacy of the colonial period." Unfortunately, all that seemed usually to follow was an African bourgeoisie's assuming places formerly occupied by colonials without demonstrating any ambition to transform the situation for the common good. Instead, they seemed more like conduits between the citizenry and outside capitalist interests while, at the same time, benefiting handsomely on their own. *Wretched of the Earth*, 152-53.

112. Martin, *Minding Their Own Business*, 114-17; Burawoy, *The Colour of Class*, 101.

113. "How the Mines Sustain Our Economy," *Mining Mirror*, October 22, 1984, 3; *Zambia's Mining Industry*, 37. The differing management styles that had originally characterized the two companies remained in place even with nationalization and the name changes.

114. Dick Hobson, "ZCCM, the Story That Began at Matero About 13 Years Ago," *Mining Mirror*, March 26, 1982, 10.

115. *Zambia's Mining Industry*, 75.

116. "Introduction to the Mining Industry for Students" (Ndola: Roan Selection Trust, 1979).

117. "Zambianization: Steps That Led to a Big Success," *Mining Mirror*, October 22, 1984, 6-7.

118. A trend that Powdermaker noticed during the 1950s and that was also documented in a 1960 study revealed that salaried mine personnel and better educated Europeans were more inclined toward working and socializing with Africans than were daily

paid white mine workers and those with less education. See Powdermaker, *Copper Town*, 69. One outside expert closely associated with the history of the Copperbelt observed that many of those in higher positions, however, were from the UK and US while daily paid laborers came more frequently from the Afrikaner population.

119. *Annual Report of the Mines Department 1969* (Lusaka: Government Printer, 1970), 1-2; Bostock and Harvey, *Economic Independence and Zambian Copper*, 124. At independence, the number of expatriate employees on the mines was approximately 7,300. Four years later, it had dropped to 4,800. *Annual Report of the Mines Department 1968*, 1. By 1970, the Zambian workforce had increased from 41,000 to 46,300, a 13% rise. "Zambianization: Steps That Led to a Big Success," 6-7. Wood records that the artisan and clerical classes, many of whom were of Afrikaans descent, were "bitterly opposed" to African Advancement and began returning to South Africa in great numbers even before independence. Many had also either lived beyond their means while in Northern Rhodesia or acquired wealth and property beyond what they could afford in South Africa, making their resentment that much stronger. *Northern Rhodesia*, 144-45.

120. "Zambianization: Steps That Led to a Big Success," 6-7.

121. One analyst recalled an instance in which the mines sent six students for study in the UK, only three of whom passed, yet the news was greeted with apparent aplomb by Zambian mine managers.

122. "Recalling the long fight for African advancement and considering the nationalist aspirations that Zambians should control their own economy and in particular the mines, the legacy of African suspicion and mistrust of the mining companies, the widely held belief that expatriate miners have a vested interest in retaining their 'lucrative' jobs, the costly business of recruiting expatriates from Europe and beyond, and the problem of a dual wage structure which follow in their wake—in view of all these factors it would appear paradoxical that the Government is so little interested in Zambianization." Burawoy, *The Colour of Class*, 99.

123. Kaplan, *Zambia*, 188.

124. Kaoma, "The Democratic Crisis in Southern Africa," 12-13.

125. As one informant with deep knowledge of the history and industry put it, "If, by increasing production you have lowered your operating costs to where you are now making money this does not necessarily mean that the shareholders will be receiving dividends/getting their money back. A lot of it you have to put straight back in as capital investment to sustain and improve future operations. . . . Mines need continual investment in new equipment and infrastructure has to be maintained so they run properly. This is generally referred to as *sustaining capital.* The mines were making money but they didn't put enough back in—they sucked it all out and things started to break down. . . . The copper price was good, the mines were working well and all they did is went and nationalized and Zambianized way too quickly with guys being stuck in positions they didn't have the experience to handle. The decline was gradual, but if you look at the statistics, it was continuous until they eventually re-privatized."

126. This was principally via the International Development Association, a subsidiary of the World Bank providing interest free loans and grants in developing countries. Total bi-lateral and multi-lateral debt was approximately $155 million USD in 1965 and high copper prices over the first few years of independence allowed the Zambian government to meet its repayment obligations while continuing moderate borrowing. By 1969, the country's debt was still only $246.6 million. It shot up to $548 the following year due primarily to external funding for the TAZARA rail line. From 1975, as Zambia's financial crisis deepened, so did its indebtedness. Kaplan, *Zambia*, 208. Kaunda was clear in drawing distinctions between foreign aid and "charity" and seemed completely

cognizant of the risks inherent in taking foreign aid. These included the rich nations' view that aid was actually charity—which continued the old patronizing relationship—and the inability, due to lack of proper training, for countries such as Zambia to survive without outside help. Kaunda, *A Humanist in Africa*, 121.

127. The accident "had a disastrous effect on production both at Mufulira and for Zambia as a whole. Production fell from an average of 617,000 tonnes per month of ore hoisted during the period May to August, 1970 to 30,000 tonnes in October, 1970." Commission of Inquiry Appointed by His Excellency, the President of the Republic of Zambia, Dr. Kenneth David Kaunda, *The Mufulira Mine Disaster: Final Report on the Causes and Circumstances of the Disaster Which Occurred at Mufulira Mine on the 25 September, 1970* (Lusaka: Government Printer, 1971), 7. Burawoy further notes, "Copper mining is inherently risky and uncertain, if only because of the technological problems involved in any mining enterprise and the unknowns that lie beneath the 'surface.' A single unexpected blow such as the Mufulira disaster can kill tens of people and put a whole mine out of action for a period of months or even years. In addition, the copper producer has to gear his production and expansion, and indeed the distribution of profits, to a fluctuating copper price sensitive to all sorts of changes all over the world which can be neither controlled nor predicted. Thirdly, the mining companies have had to be continually adjusting to the changes in the distribution of political power amongst the interested parties." *The Colour of Class*, 91-92. The importance of Mufulira to Zambia's overall economic condition has been further testified to by evidence that Rhodesian sympathizers attempted to take the mine out of commission in 1978. A foiled plot to bomb the plant arose as part of Rhodesia's effort to sew confusion among various groups participating in the southern African liberation struggle. Stiff, *See You in November*, 199ff.

128. See the work of Jubilee Research and the New Economics Foundation for detailed accounting of links between OPEC's investment decisions and multi-lateral loans encouraged upon newly independent African states, http://www.jubileeresearch.org/ and http://www.neweconomics.org (accessed April 20, 2009). See also Lishala G. Situmbeko and Jack Jones Zulu, "Zambia: Condemned to Debt. How the World Bank and IMF Have Undermined Development" (London: World Development Movement Report, 2004).

129. Situmbeko and Zulu, "Zambia: Condemned to Debt," 20.

130. The coupon subsidy program did nothing to help increase food production and was also abused. Some of the rioting that subsequently arose can be traced not just to removal of food subsidies but to general frustration with a distribution system that did not function well. Kaoma, "The Democratic Crisis in Southern Africa," 18-19.

131. Europeans had noticed problems developing with the infrastructure that was important to their way of life even earlier. Many had departed the country as a result. Stiff, *See You in November*, 86.

132. Mention of this appeared frequently in the industry's publication, *Mining Mirror*. Pertinent examples may be found in issues such as March 31, 1978, 1-3; July 27, 1979, 7; and October 26, 1979, 10-11.

133. Symptoms of Kaunda's waning popularity began to appear years earlier, particularly in voter turn out that dropped sharply from 1964 to 1972. Participation went from 94% in 1964 to 82% in 1968 and then 39% in 1972—a more than 50% drop. Kaoma, "The Democratic Crisis in Southern Africa,"12; Kaplan, *Zambia*, 39, 130.

134. Hobson, "ZCCM," 10-11.

135. Former RCM and ZCCM administrator, interview with author, 2005. Government had also increased its shareholding in NCCM from 51% to 60% in 1978. Kaplan, *Zambia*, 188.

136. Expatriates familiar with the two companies at the time seem universally to agree that RCM's executive would have made a better CEO of ZCCM, but that the CEO of NCCM was more popular with the government. Some outside observers saw an excessively formal work style in a mine setting as ultimately indicative of bad management. In imitation of British executives, those directing ZCCM would go to work in suits and ties, but "are you going to go out and see how things are really running if you're all dressed like that in your cool office?" pondered one consultant who regularly wore jeans and T-shirts.

137. Former ZCCM general manager and current senior executive, Copperbelt mining company, interview with author, 2004.

138. A *Mining Mirror* article from this time period suggests that tensions in the industry were reflected in attitudes toward foreigners. One source, in attempting to account for the condition of the mines, was quoted as saying, "There were also unsolved problems of armed robberies, black marketeering, smuggling *and influx of aliens from outside our territorial border.*" "Let's Grow More Food, Appeals Shapi," *Mining Mirror*, October 22, 1984, 1 [emphasis added].

139. According to some who participated in this effort, the consultants' presence was controversial not only among Zambian staff but for remaining expatriates who apparently saw them as threatening to their own job security.

140. I.e., price controls on multiple commodities.

141. Former economic advisor to President Kaunda, interview with author, 2005.

142. Situmbeko and Zulu, "Zambia: Condemned to Debt," 20-21.

143. This was confirmed by informants as well as Kaoma, "The Democratic Crisis in Southern Africa," 19.

144. Kaunda's government had produced other comprehensive development plan documents for the nation in the 1960s and 1970s. The program of the late 1980s did not forfeit its repayment obligations but did limit debt service payments to 10% of net export earning after required foreign exchange reserve quotas had been withheld. The government had also suspended payments to the World Bank once before. Situmbeko and Zulu, "Zambia: Condemned to Debt," 20-23, 57 and Geoffrey W. Silavwe, *Some Aspects of Personnel Management*, 7-10.

145. Situmbeko and Zulu, "Zambia: Condemned to Debt," 20-21. "We had been naughty boys," Kaunda wryly recalled of this era; interview with author, 2005.

146. The World Bank also seemed to have shifted its philosophical stance toward parastatal companies. In the 1980s, the Bank appeared to have a more lax attitude, but this changed significantly in subsequent years. Some informants also recalled this era as being the time when the privatization process really began. From the mid-1980s, for example, ZCCM had been attempting to move out of mine township service provision. Mplembe Properties was formed to manage this work.

147. Long term resident, departmental and divisional manger, interview with author, 2005.

148. His sale of township houses to miners has been seen by some analysts as Chiluba's own pre-election bid for the populace's affections. Further, it is possible that the change in government was not so much a public agitation for multi-party elections as it was simply a change from Kaunda himself. Kaoma, "The Democratic Crisis in Southern Africa," 33.

149. President Chiluba mandated this designation and it is now part of the country's Constitution. There are questions about the extent to which his decision to designate the country this way arose from theological or political convictions.

150. Former ZCCM general manager and current senior executive, Copperbelt mining company, interview with author 2004.

151. Haynes, *Religion and Development*, 8.

152. "Zambia Country Assistance Review," Operations Evaluation Department Report no. 15675 (Washington, DC: The World Bank, 1996), 18.

153. Ferguson severely criticizes the theoretical economics that influenced the types of policies that the World Bank imposed on Zambia at this stage. He specifically mentions that agricultural programs designed to turn retrenched miners into agronomists were poorly thought out in terms of land and disease conditions. *Expectations of Modernity*, 240; "Zambia Country Assistance Review," 18; Situmbeko and Zulu, "Zambia: Condemned to Debt," 23-26.

154. "Mining Industry Under New Era" (photocopy, Chamber of Mines of Zambia, Kalulushi, 2004), 10 and informant interviews with author, 2005. For a detailed history and analysis of the financial aspects of the privatization process, see Craig, "Putting Privatisation into Practice."

155. I.e., Kwame Nkrumah, first president of Ghana.

156. President Frederick Chiluba, interview with author, 2005.

157. Criticism of Zambia's over dependence on copper appears regularly in World Bank publications. See, for example, "Turning an Economy Around: The Challenge in Zambia" (OED Précis no. 132, The World Bank, Washington, DC, 1996); "Zambia Country Assistance Review:" "Zambia Country Assistance Strategy" (Country Department 3, Africa Region Report no. 27654-ZA, The World Bank, Washington, DC, 2004); and "Zambia Prospects for Sustainable Growth 1995-2005" (Economic Management and Social Policy Group, Africa Region Report no.15477-ZA, The World Bank, Washington, DC, 1996).

158. "Zambia Prospects for Sustainable Growth 1995-2005," 11.

159. By the mid-1980s, economic diversification was being stressed with some urgency, "Let's Grow More Food," 1. Kaunda himself was predicting that within 15-30 years the mines' contribution to Zambia's economy would be "insignificant." "Tough Days Ahead—KK," *Mining Mirror*, July 27, 1984, 1. "Zambia Country Assistance Strategy," 94; Donald McNeil, Jr., "For Sale: Zambia's Rich Copper Mines, All of Them," *New York Times*, August 11, 1996, http://www.nytimes.com/1996/08/11/world/for-sale-zambia-s-rich-copper-mines-all-of-them.html?pagewanted=1 (accessed January 7, 2010).

160. It would appear that this sense of pride was appealed to during the privatization process as evidenced in stories in the *Mining Mirror* about retired miners. For example, Sunday Chali, who went to work as a lasher in 1959, observed that miners of his era would listen to older and uneducated, but experienced, peers. "Retired Miner Looks Back on 35 Year Career," *Mining Mirror*, February, 1995, 11. Another story, entitled, "Sylvester Nkhoma Recalls Good Old Days in the Mine," featured a man who worked underground from 1933 to 1978. Mr. Nkhoma was cited as observing that workers in his era woke up every morning with enthusiasm and went back home happy even though safety education was "unheard of and miners survived by the grace of God. There was little barring down, no chain ladders for chute raises and accidents were the order of the day. Because of this, miners' wives and relatives used to hang around the shafts at the end of each shift concerned for the safety of their husbands and relations. 'Getting off shift was, therefore, an emotional affair.'" *Mining Mirror*, September, 1997, 7.

161. McNeil, "For Sale: Zambia's Rich Copper Mines"; President Frederick Chiluba, interview with author, 2005. The decision to privatize ZCCM was depicted in graphic terms in local publications. See "Penza Weathers MUZ Storm," *Sunday Times*, September 23, 1992.

162. In Zambia, layoffs are typically referred to as "prunings" or "retrenchments."

163. The Zambian Revenue Authority, however, pressured these same companies for payment of taxes and supply company owners sometimes found themselves in the ironic position of trying to convince one arm of government that their tax delinquency was due to non-payment by another arm of government.

164. Some informants listed the figure at $1 million USD/day. Others speculated on $500,000-$750,000/day. "Between 1996 and 1999 ZCCM made losses totaling US$1.2 billion. Despite making these losses, government continued to pump a staggering US$1 million a day in the mines, depleting the country's foreign exchange reserves in a bid to save ZCCM. In 1999 alone, according to the World Bank, Government expenditure averaged 15% of the countries [sic] GDP, almost totaling 5 years of the country's budget." Calvin Kaleyi, "Copper Crisis in Zambia," TradersAfrica.com, August-November, 2002 http://www.tradersafrica.com/articles.asp?articleid={EB470C44-E005-4111-98DE69D2-1DE0EF3F} (accessed January 7, 2010).

165. Former ZCCM general manager and current senior executive, Copperbelt mining company, in interview with author, 2004.

166. This was chronicled in public relations material put out by the mining companies and the public print media. See, for example, "Notes on the Privatization of ZCCM" (press release, ZCCM Archives, Ndola, February 1992); "Will Privatization Auction Zambia?" *Sunday Times*, September 27, 1992.

167. The financiers were principally N. M. Rothschild and Sons who were brought in, as were the attorneys for the privatization process, by the World Bank. McNeil, "For Sale: Zambia's Rich Copper Mines." According to a former ZCCM official, it was clear that Rothschild controlled the process and that the Zambian government was the client. Rothschild also had representatives on the De Beers Company board. Pallister, Stewart, and Lepper, *South Africa, Inc.*, 16.

168. From a technical and economic standpoint, Luanshya had not been profitable for years, but ZCCM's interest in job creation and maintenance seemed to have been the driving force behind keeping the mine open. It was also made clear that the new, privately held mines would not be responsible for municipal and social services. "ZCCM Transformation Plan" (Item 14.3.9E, ZCCM Archives, Ndola, September 16, 1998).

169. President Frederick Chiluba, interview with author, 2005.

170. The consortium consisted of Anglovaal out of South Africa, Noranda from Canada, Phelps Dodge from the USA, and the Commonwealth Development Corporation from Great Britain.

171. President Chiluba, interview with author, 2005.

172. Long-term resident, departmental and divisional manager, Copperbelt mining company, interview with author, 2005.

173. According to expert judgments Konkola Deeps represented, for its size and grade, one of the greatest undeveloped copper properties in the world.

174. Harry Oppenheimer was the son of Ernest Oppenheimer. Anglo also had a 27.3% share in ZCCM.

175. "Kafue Deal Collapse May Delay Privatization," *Sunday Times*, April 5, 1998, http://www.btimes.co.za/98/0405/comp/comp4.htm (accessed January 7, 2010). With the notable exception of Copperbelt Energy Corporation (CEC) proposals for Zambian share holding in the newly privatized companies appeared to be ignored or overlooked. By the mid-2000s, CEC, ZCCM's former power division, had some local shareholders but was majority owned by National Grid out of the United Kingdom and Cinergy GP from the United States. "Copperbelt Energy's Changing of Guard," *Sunday Times of Zambia*, March 5, 2006, 8; "Copper Studies," *CRU Analysis* 32, no. 11 (May, 2005): 4.

176. "Luanshya, Chibuluma Sale Pact Sealed," *Mining Mirror*, September, 1997, 1. ZCCM-IH, the mines' holdings company generally retained from 10% to 20% share in the re-privatized companies. http://www.zccm-ih.com.zm (accessed December 31, 2009).

177. *Mining Mirror*, July, 1997, 8-9. The LuSE was set up in 1994 but as of 2004, the second stage of the overall process had not been implemented.

178. Former ZCCM senior executive and current ZCCM-IH manager, interview with author, 2005.

179. See timetables of these events as rendered in Situmbeko and Zulu, "Zambia: Condemned to Debt," 24 and "Zambia Country Assistance Strategy," 59. Technical documentation indicates that, as of 31 January 2000, ZCCM had accumulated $90.7 million USD in debt to bi-lateral and multi-lateral donors. Zambia Consolidated Copper Mines Limited, "Proposed Disposal of the Assets Forming the Mufulira Division and the Nkana Mines, Concentrator and Cobalt Plant and Grant of a Second Option Over the Nkana Smelter, Refinery and Acid Plant" (photocopy, ZCCM Archives, Ndola, February 24, 2000).

180. "ZAAM Sale is Set in Motion," *Mining Mirror*, June, 1996, 1; "ZCCM Copper Lower Than Estimates," *Mining Mirror*, February, 1997, 4. Later, Kansanshi came under the ownership of First Quantum Minerals of Canada.

181. "Privatization on Course," *Mining Mirror*, December, 1997, 1.

182. "Binani Wins Bid," *Mining Mirror*, July 1997, 7. According to the Mining Mirror, Binani intended to increase copper production to 100,000 tons per year and to build an acid plant and smelter.

183. Sam Phiri, general manager of Roan Antelope in discussing that mine's sale was quoted as saying, "to some extent, we feel lost. The feeling of being detached from family cannot be avoided. It was going to be some experience to be on our own." "Luanshya, Chibuluma Sale Pact Sealed," 1.

184. The World Bank's published documents from the late 1990s praise Zambia's overall privatization program as a model of best practices for sub-Saharan Africa. Situmbeko and Zulu, "Zambia: Condemned to Debt," 26. See also the various World Bank reports cited herein. After AAC withdrew, the Bank commented that, "significant progress had been made on privatization up until the unraveling of the ZCCM privatization. By the end of 2002, a total of 254 out of the 280 parastatals under the ZPA had been privatized. The withdrawal of Anglo from Zambia, coupled with the widespread perception that privatization is largely responsible for rising unemployment, created a strong political and popular backlash against privatization." "Zambia Country Assistance Strategy," 94.

185. The price in 2002 reached a 30 year low. With the price per pound hovering around $0.67 while break-even was at least $0.70, AAC saw no reasonable way it could continue restoration of old sites or take on development of new ones. So, of its 11,000 employees, 7,000 to 8,000 would gradually be laid off or retrenched and the Nchanga open pit would continue on schedule for closure by 2010. KCM chief executive, as cited in "Enhancing Economic Diversification in Zambia's Copperbelt," Report of a Fact Finding Mission to the Copperbelt by Private Sector Donor Group, August 11-14, 2002, 7.

186. "Kafumukache: Refined Politician," *Zambia Daily Mail*, December 9, 2004, 7. See Situmbeko and Zulu, "Zambia: Condemned to Debt," for a case study of the RAMCOZ mine.

187. "Privatization: Opportunities Still Abound," *The Investor Magazine* 2, (Quarter 1, 2002): 19-22.

188. See statistics provided by Zambia Privatisation Agency/Zambia Development Agency, http://www.zpa.org.zm/ (accessed January 7, 2010).

189. See various World Bank reports cited throughout this study, especially "Zambia Country Assistance Review" and "Zambia Country Assistance Strategy," 103-05.
190. Former ZCCM-IH administrator, interview with author, 2005.

♀

Chapter Four

Private Stories of Zambia's Mining History

The World Bank's admission was correct, of course. What mattered most to Zambians largely went undetected by outside policy makers, even by committed consultants who had worked so hard on a plan that seemed infinitely sensible at the time. While economically oriented institutions of the age of development have not customarily incorporated cultural considerations into their mandates, the Copperbelt story still illustrates a point about the perils of policy making done at some distance. When we view a society or situation from a removed stance—be that physical, intellectual, or both—images of the other form before any relationship can be established. Consequently, any actions we initiate from removed viewpoints can become products of *interpretation* rather than *interaction*[1] and anything that transpires afterward may differ considerably from what might be expected. This problem can be compounded if the unexpected or different results are *themselves* interpreted at a remove. In other words, development situations that involve acting across cultures can be subject to two interpretational missteps. An initial false, or misleading, interpretation of the scene or on the ground circumstance can spawn an action that produces results other than those expected by observers. Then these results can also be misunderstood in reference to the situation under review.

In Zambia's case, for example, explanations that would accord with a dominant development narrative could be found for many of the controversies encountered during the re-privatization process. Miners who worried about their housing and felt a territorial pride in the copper industry even when it verged on collapsing could be explained in the language of organizational behavior and management: such worker sentiments often emerge in situations where more cost effective or efficient systems are introduced. But, even assuming that resistance to change is a universal human trait, does it necessarily follow that Zambians' reactions were simply another instance of this phenomenon? What if the criteria for analyzing human behavior as defined by strains of a dominant narrative had missed some considerations vital to understanding that particular setting? What if certain experiences or ideas that were integral to the history of

Zambians' existence remained wholly unaccounted for from a Western-inspired policy making purview?

The Zambian case does, in fact, suggest that cross-cultural contexts can be far richer and more complex than outside observers might suspect. Partially this may be due to the deliberate hiding of or keeping private any alternate narratives. Other research has demonstrated this phenomenon in situations where dominant ideologies attempt to control complex circumstances.[2] It can also be that outsiders simply may not know where to begin looking for substantive differences in viewpoints. If this is the case, examples from the story of Zambia's mine privatization effort suggest that a good starting point may be found in moments of puzzlement or wonder experienced by those somewhat removed from the situation.[3] Seemingly minor incidents that, in professional development parlance, would probably be termed examples of mismatched priorities or communications mix ups can take on qualitatively different connotations when we open up the range of interpretive possibilities.

For example, the manner of conducting work sometimes appeared quite different from what Western experts expected. One particularly technical case concerned how copper ore was classified. Prior to the mines' sales, ZCCM had used a classification system that divided it into "fully developed," "partly developed," and "undeveloped" reserves based on the number of drilled samples taken and on the amount of underground development already in place within the mine.[4] By contrast, systems that the World Bank consultative teams recommended were designed differently with one favored system being that of the London Stock Exchange (LSE).[5] This way of accounting for ore bodies grouped reserves into "proven" and "probable" categories determined, in part, by sampling results.[6] The system also took into account economics, market forces, and intangible commitments simply to do the mining work. What this meant was that under the LSE system there was no need to have any physical infrastructure in place at all for ore to be considered "proven" or "probable."[7] So the LSE way of understanding mine development reflected an emphasis on intangible factors, specifically on economic and monetary means of evaluation. The ZCCM system, by contrast, appeared to be more tangible and sensory in its requirements that material substance and economic projections be linked. Financiers and analysts liked the more universally recognized LSE classification system, maintaining that its adoption would be in the companies' best interests. Eventually the LSE system did begin to dominate the privatization process. But Zambian technicians, who could use either method fluently, continued to talk about the overall context by maintaining the ZCCM system for informal situations.

Outsiders remarking on this situation just noted it as a mild oddity and perhaps it was nothing more than an example of reinforced group habits of mind. Perhaps, however, it suggested something else about how many Zambians approached their material working environment. Perhaps the concreteness of construction meant something significant over against the abstractness of numerical projections. The value in entertaining this latter possibility is that it provides an opportunity for considering qualitative alternatives to predominant modes of thinking.

This study maintains that alternate narratives to the dominant discourse did exist within the Copperbelt setting. Further, it appeared that at least some elements of those narratives were grounded in traditional knowledge or experience that remained largely out of sight to non-Zambians unless deliberately sought.

This chapter begins the study's overall attempt at identifying elements of these alternative points of view but does so with an acknowledgment of the relative dearth of information available to outside observers. There are several reasons for this. One practical consideration that confronts all students of Africa is the lack of written records originating from the continent. Since local language orthographies were invented comparatively recently, most Africans have neither much of their own recorded history on which to draw nor have they generally developed the sort of reading-writing habits that have typified Western environments for some centuries.[8] This situation hampers any attempt at learning about the continent. A second potential reason for the scantiness of sources available for this study in particular concerns the possibility that Zambians were hesitant to share too much of what they knew with anyone from outside their culture. Given the experiences of past African-Western encounters this would have been a perfectly reasonable stance to take.[9]

Still it would be unwise to equate scant evidence with the idea that nothing of importance existed.[10] Indeed, that a great many important ideas were present suggested itself in snippets of literature from the colonial era on to the present day. It was obvious, for instance, that Africans in the Copperbelt region knew something about the technical skills of mining long before Europeans brought in their industrial expertise. Less obvious, however, was that they also appeared to have a certain type of knowledge of the mines and of copper itself that predated the colonialists. Further, this knowledge affected people's understanding during the re-privatization period.

Traditional Views of Copper

African cultures traditionally valued copper more than gold or silver, two metals that keenly interested Arab and European traders. Knowing that copper's malleability and inherent beauty made it more useful than these metals[11] Africans worked copper into wire forms and shaped it into tools, basins, and jewelry. Copper used decoratively also made impressive statements about the wearer's wealth, status, and leadership abilities.[12] And, as a medium of exchange in Central Africa, cross-shaped copper bars were widely available along trade routes throughout the sub-Saharan region.[13]

None of this information represents newly discovered antiquarian detail. But what may be news within the development policy making arena is that precolonial Africans chose their materials not simply by taste or convenience but with deliberate reason. Copper physically expressed deep cultural values or beliefs that ventured beyond economic concerns into religious and spiritual realms.[14] Diviners used copper; Africans guarding against witchcraft used it as well.[15] In Zambia, copper—the Bemba word for which is *umukuba* or *mukuba*—

was also symbolically bound to the wider physical creation through an association with lightning. The Copperbelt's spectacular thunderstorms continue to testify to this powerful connection.

In traditional society, religious connotations also infused copper production processes and the people involved with those processes. Before mining proceeded, local chiefs and workers would pray to ancestral spirits who presided over the mines, being especially careful to plan excavations that maintained harmony between the technical work of digging and the earth being dug.[16] Ancient smelting practices were regarded as mysteriously sacred endeavors and the African artisans who could transform one substance into another, as if melting stones, were especially revered.[17]

Although historical information is scant concerning the ways that Africans' religious understandings of copper affected their economic activities and vice versa, knowledge of the mining environment's overall value to Africans is a bit easier to come by. We have already seen, for example, evidence of a passionate protectiveness by Zambians toward the mines during the privatization process. Observers have also for decades noted how underground mines appeared to evoke the African consciousness in ways similar to on-going debates about land rights.[18] In both cases—agricultural land and underground mines—collective ownership models traditionally prevailed and a palpable sense of relationship existed between workers and the physical setting itself.

Not all of this experiential background has escaped the notice of development policy makers and analysts, but how it has been interpreted bears comment. The World Bank, for example, recognizing that Zambians were protective of their major industry, sometimes referred to their attitudes as exhibiting an "enthrallment" with copper.[19] Sarcasm may not have been the Bank personnel's explicit aim in such remarks. But, from a standpoint that takes seriously both Rhodes's original exploitive intentions for the region and Zambians' traditional history, it is an intriguing word choice. Additionally, numerous outside observers have commented on the collectivity and relationality that traditionally bound Africans to their lands, a relationality that was still evident in Zambia during the mid-2000s.[20] Exploring how these attitudes possibly reflected religious values that, in turn, provided commentary on development policy making has not been fully addressed heretofore. Locally told stories of Zambia's mining history illustrate how rich a field of phenomenological exploration the Zambian experience could be.

Traditional Copperbelt Guardians of Creation

In traditional Copperbelt life beings with supernatural powers guarded various regions of creation.[21] These differed from region to region according to what the local people revered but in all cases the guardians rewarded those who treated society well and punished those who did not.[22] The power on land was sometimes understood to be a lion. The power associated with water was frequently likened to a mermaid and the power of the underground was said to be a snake.

On land, if someone wronged or carried a grudge against another person and then encountered the guardian lion, he or she might well be consumed on the spot. If, however, one had a clear conscience toward the rest of the community, an encounter with the lion might bring rewards. In such cases, the land's supernatural guardian was known to kill fresh game and leave it for well-deserving individuals.

By contemporary times, the mermaid guardian of the waters throughout major sections of Zambia had come to be known as the donafish. Whatever might have been the original terminology for this supernatural guardian, her Romance language inspired name probably came from the mixing of Italian and African laborers during the earliest days of the colonial project.[23] Popular contemporary understanding of her as a beautiful, well-endowed white woman was likely linked to that time as well and, in the mid-2000s, images of the donafish, embellished by jewelry and wristwatches,[24] occasionally adorned the stalls of traditional healers in local townships.

As with the lion guardian, the donafish could leave something of value near a body of water and those exhibiting enough courage to retrieve it would become wealthy.[25] Back in the colonial days, stories about the donafish encouraged men to come from outside the Copperbelt to work in Northern Rhodesia's mines. With those mines being some of the wettest in the world, surely the donafish should be nearby. After independence, local school children were known to linger near ponds or streams hoping to spot the mermaid or at least any jewels or money she might fling onto the banks. Donafish sightings were also occasionally made in the 2000s. One was said to be kept by a man in a swampy area near Lusaka, for instance. Residents eventually sought traditional medicine to kill the man and, when he died, the donafish soon died as well.

Perhaps even more than the donafish, the underground's snake guardian was legendary in certain Copperbelt regions with magical snakes continuing to appear in various locales during the re-privatization era. Zambian culture traditionally associated snake imagery with maleness and Zambian miners, like miners everywhere, enjoyed renown for their bravery and strength.[26] On the Copperbelt, the Lamba word for "snake" was the same as the word for "chief" signifying that reverence was due it.[27] While snakes understandably could be frightening, their presence, similar to that of the lion and donafish, guided worthy people to greater things. In the mining environment, big green snakes might pass by shaking the ground like an earthquake. Those who saw them were advised not to be afraid but to mark the snake's location and dig for copper there. The snake guardian's passage or presence near mining endeavors was also considered crucial for protection. So laborers who got mixed-up with witchcraft or failed to demonstrate proper respect prior to commencing work risked alienating the snake, thereby exacerbating the dangers already inherent in their perilous tasks. The nephew of a local chief explained the phenomenon this way:

Nephew: [E]ko icili iyi tuntulu, icili ilepita, ilapita cila mwaka.

Nephew: [T]here is a living one, which is still passing, it passes every year.

Interviewer: Yamusango nshi iyi insoka?

Interviewer: What type of a snake is it?

Nephew: Iyo insoka ilapita panshi kwati ni earthquake. Ilapita panshi nga yapita panshi, nomba nga kwacilamo kulya kwafula abakweba ati imitima yabo ya bufwiti, ne mano yabo tabalingile ukubombela kulya nga taipitile insoka, ninshi amasanso kuti yatatika ukusangwa kumine. Kulaba limbi filya ati nga taipitile ilye nsoka ba amba ukulafwa bamo bamo filya balefwa limbi cimakalota camushika mu matipa, limbi driver talemonako, limbi engile pamugodi caputuke icibumba ca mwilikisha. Nga baya mukupupilila pupilila filya fine ba pepelela ifimupashi iya cilya cine, elyo chief nabebo babeta, nga basuminisha ukwebati konkanyenipo kubomba, nga bakonkanyopo nacena ninshi cayamba ukupita, ico cine, capita caisa mukupempula nomba, oh uku balemfwaya, capempula caya. Nomba ninshi nomba kutampo kubomba ninshi pawama nomba, na maeligo nabena, nabo nga capita ico, nabo elyo bakwata sana indalama, mulandu wakweba ati, abene baselula amabwe.

Nephew: That snake passes underground like an earthquake, it passes underground. When it passes, miners can work properly. However, if miners are highly wicked due to witchcraft, snake would not pass. Miners are not supposed to work before the snake passes, or else many accidents are likely to occur. Sometimes, if the snake does not pass, some people start dying, at times the machine would bury one in the mud, at times the driver would fail to see, maybe when he is underground the tunnel collapses on him. When they go to appease in accordance with the Lamba tradition and the Chief also gives consent to continue working, and when they start, then the snake would pass. When it passes, it comes to check on the miners saying that "people want me" and goes back. Afterward, they continue to work in peace since the mine has been cleansed. When it passes even the illegal miners also make a lot of money because the owners have brought the stones to the surface.[28]

Floods, accidents, and mysterious deaths brought on by a simple exhalation of the offended guardian's breath would surely follow any irreverent acts. To avoid such fates, people customarily offered gifts such as cartons of soap, mealie meal, red powder, or even several meters of white cloth. These items would be thrown into a river or other area accessible to the snake guardian alone.[29]

In the mid-2000s, Zambians understood that supernatural snakes still played an active role in their society.

Mr. N.: So efyo caba, caba nga fino fine tawlikuno kwali icisoka ico beta ati, 'Chipimbi.' According to the Lambaland Chipimbi ni nsoka iyo babika ati nimfumu, so nga mwalufyanya kuli Chipimbi ninshi anything can happen pamugodi. So ema belief yabako. So ilingiline ngamulefwaya ifya kukuula nangu ni church, ificitika sana mukasanga pali icisoka icikulu sana.

Mr. N.: So that's how it is, just like we have been here there was (is) a big snake called Chipimbi. According to the Lambaland Chipimbi is regarded as a Chief, so if and when you wrong Chipimbi, then anything can happen on the mine. So these are the beliefs that exist. So most of the time when you want things to do with construction and building even things like a church, you usually find a huge snake.

Abene muno mwilamba ngabamona filya kuya kumfumu ba Mushili nabambi, ukuyamukutota. Batwala ifya bupe, ubunga, inkula bayatota kulya. Nangu mulefwaya ukutampa maine nabasungu balacita bow to that. Balabeba ati, paku tampa iyi mine kano imfumu ya sumina, mwaipela fyakuti, so kulaba ukupele mfumu portion imo. They are entitled to get something within the mining, otherwise ngatakuli ifyo ninshi ilye nsoka ileisa. Pa maine limbi umugodi wa flooder, na ma accidents ninshi mwaishiba. So there is a relationship traditionally ukulingana nentambi because those are spirits, so they used to worship them previously and they are carried on and on to anything existing in that environment. So namufya mining, those snakes are counted ukutila they are important.

Owners of Lambaland when they see that, they go to chief Mushili and others to go and appease, give thanks and gifts such as maize meal and red powder. Even when you want to start a mine, even *bazungu*[30] bow to that. They are told that for them to start a mine, the chief should agree and you give the chief such and such a token and they do just that. They[31] are entitled to get something within the mining, otherwise without such tokens of appreciation snakes go on the mine and as such there are floods and other accidents. So there is a relationship traditionally between snakes and mines because those are spirits, so they used to worship them previously and they are carried on and on to anything existing in that environment. So even concerning the mines and mining snakes are counted as important.[32]

All this sometimes seemed to conflict with Western Christian values and could cause problems on a number of fronts.

Mr. N.: So na nomba line because of that belief insoka shilacita operate muli ilya ine inshila, filya fine abantu basuminamo. Kabili that's how nga kukonka noma ifi ifya kuchurch, that's how we say the Devil works because there he is also threatening people to say "I also exist." Pantu ngamwasumina, he can perform wonders. Tefyo balanda ati nabo bene bakalacita perform ama miracles. So the

Mr. N.: So even now, because of the belief, snakes operate in that belief system of the Lambas. If we follow the Church things, that's how we say the Devil works because there he is also threatening people to say "I also exist." Because if you believe, he can perform wonders. Is not what they say that he'll be performing miracles. So the devil performs wonders in those beliefs so that he can be feared. So devil performs wonders in those beliefs icakweba ati naine balentina ati eko naba. So eko fyaba elyo filacitika and they will continue to happen. Nalilya twali naba Kasanga pakutampa church ilya iya ku Ashford, even up to now there is a very big snake which comes there.

these things are there and they will continue to happen. Even when I was with Mr. Kasanga, when we were starting to build that church in Ashford, even up to now there is a very big snake which comes there.

So balanda ati kano mukalambe kumfumu mukayelanda ati palibe nsoka apa. Bene bacalo, ba Mushili abaLamba abene bali kulya. Mukachite nani kulya, mukaye kwiLamba kuli baMushili mukayelanda fyakuti fyakuti, aaah. It's a problem to us to do that, tulacisanga filya fine cafumapo cabutuka caya, mwaisayako limbi limbi mwaya cisanga muchurch from nowhere mwacisanga nacikalafye mucorner, lulya mulelanda so, efyo fiya! Muletonkanya ifyakufwaya icimuti, disappears just like nothing icakwebati tamwishibe napo cafumina. It's there, kulya kuchurch kulya kuAshford.

So they say that the snake will only go when you pay homage to the chief and say there is a snake here. It is the owner of the country, Mushili, a Lamba who is down there. They have told us to go to Chief Mushili and to do such and such, aaah. It's a problem to us to do that, we find it but it vanishes just like that. You go another time you find it in the church. From nowhere, you find it in the corner of the church. As you are speaking, it vanishes! Just when you are thinking of finding a stick to use to kill it, disappears just like nothing such that you can hardly tell how it came out. It's there at the church in Ashford.[33]

Indeed, experience of the underground snake guardian was especially palpable in Luanshya, the Copperbelt town that was also subjected to early vagaries of the re-privatization program.

"These Things are There."

Luanshya was popularly considered one of the two most attractive towns on the Copperbelt and everyone's best romantic impulses found something appealing about the industrial era origins of its mine. Europeans interested in Zambia's mining history loved to recount how prospector William Collier shot a roan antelope that fell on a copper-stained rock near the Luanshya stream in 1902. Zambians nuanced the story by noting that, after it fell, the antelope's horns pointed to the copper deposit. By either interpretation, the roan antelope was significant to the region and for much of its history the industrial mine built there was known as Roan or Roan Antelope.

Luanshya's association with the roan antelope came in conjunction with extended snake difficulties, however. Mine workers recalled their fathers telling stories of a big snake that would pass by, sometimes at the peril of local people who died following encounters with it. The best protection against it seemed to be appeasement involving both the local chiefs and prayers to the workers' ancestors. Even so, everyone lived with constant worry because the snake was still present, lurking and occasionally lashing out.

Depending upon whose version of the story was being told the underground snake had different names and a variety of misfortunes that the town experienced had diverse causes related in some way to the guardian. One famous story reached back to the early days of prospecting and industrial mine establishment when there was a high incidence of death within the local population. Africans largely assumed this misfortune signaled something about the snake guardian—whom they knew as Chipimbi—and his requirements for respectful treatment. As far as Europeans were concerned, the local people simply feared a many-headed snake named Sanguni but the deaths were due to malaria. Since the problem persisted, the mining company eventually commissioned Chirupula Stephenson, a European married to the daughter of a Lala chief, to do an exorcism. That event was widely recorded and generally termed a success.[34] But what really made it so, according to Europeans, was that the industrial mining company engaged in a parallel effort to clean up the Luanshya stream, draining stagnant swamps where malaria infested mosquitoes lived. Zambians, however, knew that Chirupula Stephenson's name meant "whipper"[35] and he had a reputation for doing just that to the local people. So there was some doubt about the sincerity and authenticity of his effort. More importantly, the area near what had become the mining town was traditionally the Lamba people's shrine to the snake. Draining the swamps in an effort to stamp out malaria was not so much a solution to the problem as it was a precursor to it. For, when the snake's habitat was destroyed, it struck back.

What *is* There?

Certainly students of culture and religion would find much of interest in tales of supernatural guardians and apparent syncretism between magical snakes and pernicious Satanic figures. But when traditional ways of thinking become present in, even help form interpretations of, circumstances devoted to modernization and progress how should they be accounted for? Even the little bit of alternate story material presented here suggests an intriguing possibility. That is: many Zambians appeared regularly to include supernatural experience as a category of thinking that had no counterpart within a dominant development narrative. In other words, Zambians appeared often to make sense of the modern, industrial mine setting by incorporating criteria for which there would simply be no place within the development policy making context.[36] In the Luanshya story, especially, there seemed to be more layers of experiences and different types of phenomena present as compared with that found in official corporate policies. There was also a sense in which the basic elements of the world seemed simply more substantive and vibrant than Western logic could grasp or the Zambian narrative wished to reveal.[37] These possibilities would suggest the need for policy makers to broaden substantively the overall parameters of development discourse. At the very least, such broadening would include deliberately seeking engagement with analysts and practitioners outside of the economic and managerial disciplines that currently hold a disproportionate influence over the intellectual foundations of the development industry.[38]

In addition to being broadened, the development discourse might also need different organization. This is because, as the hidden story starting to be glimpsed here illustrates, information from the local context can come to outside observers more as a collection of snapshots than as a clearly drawn, logically related narrative. Consequently, coming to greater appreciation of how alternative narratives inter-relate may require adherents of dominant viewpoints to leave certain elements of logic behind. Outsiders might not know how to arrange all the elements of such an alternative narrative. But, fortunately, that is not always the most important task. Simply being aware that alternate or private stories exist can be a monumental breakthrough when development policy making is designed. The next three chapters will attempt to articulate some significant qualities pertinent to several experiential aspects of Zambia's mine re-privatization project.

Notes

1. Alfred Schutz, "The Stranger," in *Collected Papers, vol. 2, Studies in Social Theory*, ed. Arvid Brodersen (The Hague: Martinus Nijhoff, 1976), 98-99.

2. Scott, *Weapons of the Weak*, 140, 234.

3. An American diplomat in Zambia once remarked that certain cultural oddities might be observed and then simply filed away in one's mind as if constituting a sort of Ripley's *Believe it or Not* until such time, if ever, as one might be confronted with the ramifications of those "oddities."

4. Zambia Consolidated Copper Mines Limited, "Proposed Disposal of the Assets Forming the Mufulira Division and the Nkana Mines," 63. Technical experts suspected that the system—apparently unique in the mining world—was designed by expatriates during the colonial era and subsequently taken up by Zambians as they rose into the highly skilled ranks during the 1970s and 1980s.

5. Examples of other classification systems are SAMREC, or South African Code for Reporting of Mineral Resources and Mineral Reserves, and the Australian system, JORC, standing for Joint Ore Reserves Committee.

6. Zambia Consolidated Copper Mines Limited, "Proposed Disposal of the Assets Forming the Mufulira Division and the Nkana Mines," 62.

7. The "proven" and "probable" categories relate to confidence levels with "proven" equating to a high degree of confidence that the ore is present and economically feasible to recover. Zambia's copper ore deposits are of a good continuity and have a low level of grade variability. So Zambians could make projections about future deposits by using comparatively few drilling samples.

8. Archbishop Robert Sarah, *Culture, Democracy, and Development in the Light of Centesemus Annus* (Nairobi: Paulines Publications Africa, 2000), 36.

9. Georg Simmel, *The Sociology of Georg Simmel*, trans. and ed. Kurt H. Wolff (Glencoe, IL: The Free Press, 1950), 330-47. Scott, *Weapons of the Weak*, 284.

10. Chinweizu, *Decolonizing the African Mind*; Reed, *Development Matters*.

11. Herbert, *Red Gold of Africa*, 70, 297; "Copper and the Copper Mining Industry of Zambia," 1.

12. *Zambia's Mining Industry*, 14-16; Herbert, *Red Gold of Africa*, 242-45.

13. Bradley, *Copper Venture*, 38; *Zambia's Mining Industry*, 16.
14. Herbert, *Red Gold of Africa*, 209-10, 241.
15. Herbert, *Red Gold of Africa*, 261, 302.
16. Cline, "The Sources of Metals," 303.
17. Herbert, *Red Gold of Africa*, 4, 33, 193. Fire was also considered a sacred symbol of morality. Lane, *Moment of Encounter*, 32. Relating what made the traditional smelting process so mysterious a colonial era missionary is said to have remarked, "stones are not things which melt." As quoted in Bradley, *Copper Venture*, 36-37.
18. Powdermaker, *Copper Town*, 64; Epstein, *Scenes from African Urban Life*, 184. See also Mbiti, *African Religions and Philosophy*, and Tsenay Serequeberhan, "Colonialism and the Colonized: Violence and Counter-Violence," in *African Philosophy: An Anthology*, ed. E. Chukwudi Eze (Malden, MA: Blackwell Publishers, 1998), 234-54.
19. "Zambia Prospects for Sustainable Growth 1995-2005," 18, 21, 32.
20. Roberts, *A History of Zambia*, 209; Powdermaker, *Copper Town*, 63-64.
21. For much of this background, I am indebted to a Kitwe folklorist and local radio personality whose uncle was the renowned Zambian author, Stephen Mpashi. The notion of regional guardians as explained by the folklorist appears to be compatible with the concept of territorial cults that Binsbergen observed throughout Central Africa in general and Zambia in particular. W. M. J. van Binsbergen, "Explorations in the History and Sociology of Territorial Cults in Zambia" in *Guardians of the Land*, ed. J. M. Schoffeleers (Gwelo: Mambo Press, 1978), 47-88. The traditional guardians in Zambian life as described here also seem exemplary of Leo Apostel's claim that pre-colonial societies generally had masters of the mines, land, and water. *African Philosophy: Myth or Reality?* (Gent, Belgium: Scientific Publishers, 1981), 292.
22. See also Hall's section on "legends of the tribes" in which he maintains that all of the approximately 70 groupings in Zambia had their own traditional stories that were being forgotten in the modern era. *Zambia*, 11-14.
23. Although the name and characteristics of the donafish are obviously highly Europeanized, evidence of a mermaid or water spirit can be found throughout Southern African regions. Ellis and Ter Haar, *Worlds of Power*, 49. The image may have been imported into traditional African society or layered upon pre-existing mythologies. There is, for instance, a primordial sister in one Bantu genesis myth who finds copper ornaments in the Kasai River and then disappears into it herself. Herbert, *Red Gold of Africa*, 287. Ellis and Ter Haar also recount a story about the underworld as being populated by spirits of those who, in life, were controlled by fallen angels. *Worlds of Power*, 49-52. Dr. Bush, a traditional healer in a Kitwe area township adorned his stall with a large donafish painting. He explained that a number of angels came down from heaven long ago and then turned into different beings including the donafish. Since these angels were originally white, the donafish is white as well. It is unclear whether Dr. Bush's story was in the same vein as Ellis and Ter Haar's but many elements seem appropriate to the Copperbelt setting. There was also considerable mixing of European and African languages throughout various regions of Zambia. Lagus, specifically notes that signage near the Kariba Dam was done in English, Italian, and "Bantu." *Operation Noah*, 41.
24. The presence of pens and wristwatches has constituted one of the earliest associations with the idea of modernity and development. Hiebert, *Transforming Worldviews*, 243.
25. Water as constituting entrance to the spirit world is a widespread theme in Central, Western, and Southern Africa. See also a discussion of the water spirit Mammy Wata who, in vast regions along the Atlantic coast, was understood to bring great wealth as well. Ellis and Ter Haar, *Worlds of Power*, 52, 121-22.

26. It may not be coincidental that copper is traditionally associated with femaleness in African cultures that have gendered nouns in their vernacular languages. Further, the alchemical sign for copper is modified from the Egyptian hieroglyph for eternal life a symbol that, in ancient Greco-Roman culture, became associated with Venus/Aphrodite. Herbert, *Red Gold of Africa*, 288; "Zambia's Mining Industry," 6. See also Mbiti, *African Religions and Philosophy*, 196 and Merlin Stone, *When God Was a Woman* (New York: Harcourt Brace Jovanovich, 1973), 198.

27. Since the Copperbelt was traditionally Lambaland, these views would have been quite pervasive in the area prior to the influx of other groups coming to work on the mines.

28. Interview with research assistant, 2005. The translator notes that the word for "owners" (*abene*) connotes spirits, not economic investors or modern industrial managers.

29. See similar ceremonies in southern regions of Northern Rhodesia as recounted by colonial observer A. H. Gibbons, *Africa from South to North Through Marotseland*, vol. 1 (London: John Lane, 1904), 45-46.

30. I.e., whites.

31. I.e., the chiefs.

32. Interview with author and research assistants, 2005.

33. Interview with author and research assistants, 2005. This complex association of snakes, traditional rulers, and Christianity calls to mind an incident related to the colonial era Watch Tower witch-finding episodes as recounted by Taylor and Lehmann. They were able to interview Zambians who had participated in the activities of Tomo Nyirenda, also known as Mwanalesa (God's child), who taught that Europeans and leaders of the organized Churches were snakes. *Christians of the Copperbelt*, 26.

34. A version of the above story was adapted by a European life long Luanshya resident from a write up by the late Dick Hobson, a long time Zambian resident whose journalistic efforts are cited elsewhere in this study and whose historical essays continue to be used in promotional literature by the country's tourism and business industries. There are also references to Stephenson's exorcism in P. M. Mukula compiler, *Calendars of the District Notebooks (Copperbelt Province) 1899-1964* (Lusaka: Government Printer, 1981), ix and in *Zambia's Mining Industry*, 32.

35. From the Bemba verb *ukulupula* meaning "to whip." Fields translates Stephenson's nickname as "the flogger," attributing it to his work as official executioner. *Revival and Rebellion*, 176.

36. By way of comparison, see Deneulin's reference to a 1999 study of Nigerian resettlement issues after a dam's construction had disrupted a local community. The community needed to offer an animal sacrifice if its shrine, a rock, had to be moved. "The resettlement authorities did not offer compensation for the relocation costs on the ground that 'they were not compensating rocks.'" *Religion in Development*, 44.

37. Taylor and Lehmann's study documents a similar phenomenon specifically within the context of African Christian spirituality on the Copperbelt. A "this-worldliness," they note, is expressed through "reluctance to divide experience into separate realms of sacred and secular, spiritual and physical." Elsewhere, in a discussion of African interpretations of development in a materialist sense, they write that, "a great deal of African 'materialism' is actually accompanied by the most lively sense of a spiritual world. It is not in fact at all what it seems to Western eyes." *Christians of the Copperbelt*, 121, 275.

38. Green cites a "lack of multi-disciplinary literacy" within the development industry, saying that practitioners hamper their effectiveness by relying so heavily upon the discipline of economics. *Poverty to Power*, 442.

♀

Chapter Five

"The Spirits Are Not Happy": How Zambians Knew Things Were Not Well

Private ownership of an organization or institution, so the theory goes, has the advantage of instilling a sense of responsibility in those involved. Responsible actions, in turn, should bring about higher quality results that eventually benefit all who participate in the system. A perennial test of this chain of logical reasoning is the issue of job safety in industrial mining. Which set of responsibilities should prevail as the most significant—the responsibility to keep up the production process or the responsibility to keep safe those who are engaged in production—has formed the basis for endless labor-management disputes.

In Zambia, as the re-privatization process unfolded, dominant policy-related talk focused on economic indicators with arguments in support of the process pointing to the benefits of job creation and GDP growth. Curiously, for men and women in the streets neither the mines' employment statistics nor their personal susceptibility to retrenchment as the companies honed their workforces appeared to be the root concern. Of course Zambians were quite worried about their economic futures but what seemed to be a deeper (albeit related) disquiet was a generally pervasive sense that the mines were becoming less safe places. In fact, the average Zambian seemed to think that, under privatization, the number of mining accidents had dramatically increased. From a development policy point of view this could seem odd, especially given an understanding that the private companies were making significant investments in order to upgrade neglected infrastructure.

Zambians, however, offered multi-layered reasons for their point of view that keenly dissected corporate policy and raised other concerns as well. As with the varying narratives of the country's mining history, debates about mine safety and accident rates had public and private dimensions. There were some obvious reasons why accidents were on the rise, Zambians said. One was that production pressures made people work too quickly; another was the manner in which operations were being carried out. As to the first, Zambians asserted that greed inspired the companies to emphasize production so strongly that further concerns became ancillary. In the past, for example, the nationalized company regularly issued and routinely replaced different colored overalls and hardhats for miners to wear. Under privatization, though, everyone seemed just to wear the same nondescript outfit until it looked shoddy.[1] In the same way, underground infrastructure such as ventilation was old and needed repair. In some places the air was so poor that cloths wrapped around workers' mouths could turn black in fifteen minutes. So neither how the miners were outfitted nor how the mines were furnished mattered in comparison with keeping production going apace even if this was done when both the miners and the mine did not appear ready for work.

Table 5.1. Persons Fatally Injured and Fatality Rates per 1,000 Men at Work 1956 to 2005; Persons Seriously Injured and Serious Injury Rates per 1,000 Men at Work 1964 to 2005[i]

Year	*Average Number Men at Work*	*Fatally Injured Number*	*Fatally Injured Rate per 1,000 men at work*	*Seriously Injured Number*	*Seriously Injured Rate per 1,000 men at work*
1956	48,476	59	1.22	–	–
1957	50,160	63	1.26	–	–
1958	39,966	36	0.90	–	–
1959	44,611	43	0.96	–	–
1960	46,117	58	1.26	–	–
1961	45,352	55	1.21	–	–
1962	41,754	51	1.22	–	–
1963	42,824	27	0.63	–	–
1964	44,718	43	0.96	3,426	76.61
1965	46,975	26	0.55	3,598	76.59
1966	45,935	36	0.78	4,067	88.54
1967	48,482	37	0.76	3,641	75.10
1968	48,611	39	0.80	3,034	62.41
1969	55,279	56	1.01	3,388	61.29
1970[ii]	56,093	160	2.85	2,897	51.65
1971	56,762	53	0.93	2,705	47.66
1972	58,267	48	0.82	2,780	47.71
1973	60,205	48	0.80	2,675	44.43
1974	64,027	39	0.61	2,775	43.34
1975	62,963	52	0.83	2,461	39.09
1976	60,520	38	0.63	2,300	38.00
1977	61,876	33	0.53	2,283	36.90
1978	58,365	33	0.57	2,604	44.62
1979	57,989	29	0.50	2,571	44.34
1980	59,175	41	0.69	2,676	45.22

[i] Source: Mines Safety Department Annual Reports. These were unavailable after 1987; figures from that point onward were obtained from the Ministry of Mines.
[ii] Exclusive of the Mufulira Disaster totals, the fatality rates read as follows: Number: 71; Rate per 1,000 men at work: 1.27.

Year	*Average Number Men at Work*	*Fatally Injured Number*	*Fatally Injured Rate per 1,000 Men at Work*	*Seriously Injured Number*	*Seriously Injured Rate per 1,000 Men at Work*
1981	58,937	31	0.53	2,155	36.56
1982	56,769	30	0.53	1,867	32.89
1983	53,339	20	0.37	1,743	32.68
1984	55,127	22	0.40	1,858	33.70
1985	53,914	39	0.72	1,778	32.98
1986	52,682	24	0.46	1,561	29.63
1987		21			
1988		22			
1989		27			
1990		26			
1991		21			
1992		24			
1993		22			
1994		25			
1995		19			
1996		25			
1997		20			
1998		22			
1999		10			
2000		8			
2001		23			
2002		24			
2003		18			
2005		80			

Table 5.2. Mine Safety Performance Statistics 1995 to 2004[iii]

	1995	**1996**	**1997**	**1998**	**1999**	**2000**	**2001**	**2002**	**2003**	**2004**
Employees ('000)	27.0	26.0	25.1	23.8	20.4	24.8	26.7	27.4	26.7	25.5
(LTI) Lost time due to injury events	850	770	803	747	532	536	426	457	448	385
(LTIFR) Lost time due to injury frequency rate	2.26	2.0	1.93	2.07	1.39	1.90	1.40	1.46	1.47	.132
Fatalities	15	18	8	17	10	6	21	14	11	14

iii Source: Chamber of Mines, Kalulushi.

Additionally, Zambians noted problems created by the ways the companies were mining. A number of miners speculated that the companies were using methods inappropriate for the Copperbelt's particular ore bodies. Why they would do so either had to be due to mismanagement or misplaced priorities. Mismanagement occurred because greedy foreign investors brought in people who really knew nothing about mining. Rather than paying local people who knew the area, they wanted their own "experts." But these experts really did not know what they were doing. The companies also demonstrated misplaced priorities by valuing efficiency over the ore body itself. How could they simply abandon whole underground areas rather than take the time to extract carefully what remained?

Instead of thoughtfully removing the ore, the new mining companies seemed to Zambians so intent upon getting what they wanted that they left only devastation and ugliness behind. As a greedy baby might carelessly stuff its mouth, chewing and grasping all at once, so the companies exhibited blatant haste to obtain every bit of ore as quickly as possible. In the process they were neglecting safety procedures, their workers' welfare, and destroying the earth itself.[2]

Less obvious reasons for the increased accident rates took Zambian speculation in a direction that outside policy makers might hardly anticipate.

"There's no Such Thing as a Natural Death."

On the Copperbelt there was a concept known in Bemba as *chisomo* or *ichisomo*. Other regions of Zambia had similar ideas sometimes named as in the Chokwe *kumumbilako* and sometimes known only by descriptive stories and teachings. Copperbelt area Zambians in their mid-twenties and older had learned about *chisomo* formally through direct instruction by their elders and/or by picking it up informally from playmates and casual social interactions.[3] The idea of *chisomo* taught that evil spirits could prompt people to do destructive things. Such spirits would not come to just anyone but would be sent to specific individuals in much the same way that an evil spirit was said to have attacked the Biblical figure of King Saul in his periodic rages against the shepherd, David.

At least two situations could elicit *chisomo*. First would be any individual death. Death was not an isolated occurrence but something that affected family members and friends in ways beyond the emotional or psychological. Through supernatural forces, the death of a person could suddenly take control of a close relative's physical actions whether or not that person was aware of being newly bereaved. Because *chisomo* could attack without warning, those who had just lost someone were considered to be in very vulnerable positions. In the early 2000s, for example, the wife of a senior level administrator at a Copperbelt educational institution died while he was in a remote part of the Northern Province. There was no way to communicate this sad news by telephone so a staff driver went to fetch him. The story subsequently went around campus that the driver madly rushed the distance to Mbala, anxious to retrieve the administrator before anything else bad could happen to him. Even without knowing his wife had died, the administrator was the potential target of a spirit that could attack, causing him to harm himself or someone else.

A second situation in which *chisomo* might operate involved bad relationships between people. One person's jealousy of another or punishment for errors committed against someone else might bring on such attacks, generally recognizable by their suddenness and unpredictability. What those from Western backgrounds might term "freak" accidents would, for Zambians, be attributable to *chisomo*. Underground, *chisomo* might be the determining factor when something untoward happened in a region where a fatality had previously occurred. Official reports of such workplace incidents could record the reason as being due to workers taking short cuts or engaging in sloppy performance. But miners themselves privately worried that the deceased's fellow workers had disliked or begrudged him. If that were the case, then the departed miner's spirit, lingering near the scene of his death, had actually reached out to get others to join him.

> [T]here are the logical reasons why things happen and then there are those things which, "I won't talk about it but I'm sure there is something that we have not been doing that has led to this." Forget about "this was an unsafe act" or "unsafe practices" and so on . . . there's always something behind. There's no such thing as a natural death or a natural illness.[4]

Sometimes technical constraints limited access to ore bodies, meaning there were occasions when work would have to continue even in areas where fatalities had occurred. Miners often felt quite uncomfortable being in such places. Sometimes they developed illnesses or maladies that would provide legitimate corporate cover for avoiding the dreaded site. Any future mishaps there would be blamed on the initial fatality.[5]

> [I]f, for example, somebody's disaffected or you know for one reason or another people don't like him and then something happens they would, you know, there's a tendency to blame him. That "perhaps he went because we don't like him." . . . You know, the relation between safe working practices and accidents, although it is being taught and you have, at the change of shift, there's always a safety pep talk where they talk about issues related to safety—five, ten minutes or so. "This area's not safe. We need to do this. We need to do this." And so on. That has been with a view of trying to get each individual to take responsibility for their own individual safety and for the safety of the section they are working in.
>
> But still when something happens it does not happen because they took short cuts. It happens because the guy who worked there before, and maybe he died in an accident or whatever, he may be calling others to join him or something like that. But . . . following a fatality [if a miner's] conscience is clear he had no ill feeling towards that person he'll go back and work. If his conscience is not clear, if he had some ill feeling against that guy, you'll find him come, you know, "I've got a back ache, I've this and this" for a good two, three months. Or he may even wish to be changed from that particular section. Because he feels that that person's spirit will come and visit him.[6]

A physical place associated with death could also elicit more death as was seen in an incident that a Copperbelt mine manager encountered while working at a South African open cast mine. When workers pushed overburden away from the high wall a bulldozer driver simply went too far and dropped about thirty meters into the pit. He was trapped and drowned in water at the bottom of the mine. During the investigation that followed this terrible incident it was discovered that the bulldozer's tracks ran through an old graveyard. Normally, when mining companies encounter a graveyard, they are expected to follow specific procedures of exhuming the graves but no one had known about this old site. So, according to laborers there, the man in the bulldozer had died not because of a procedural problem or operator error but because the old graveyard had been disturbed.

Another idea closely related to *chisomo* was *iminyama*, something considered worse than bad luck. Like an evil aura, *iminyama* would attach itself to a person and go wherever that person went. Ultimately, it could cause people to do strange or bizarre things and, like *chisomo*, could select victims regardless of their physical proximity or awareness. This possibility carried special and ironic import for the mines where *iminyama* had the potential of harming laborers who were already participating in hazardous activities. Underground, miners were physically cut off from their families for their entire work days making them—it would seem—totally removed from what was happening on surface. In other settings such arrangements might be considered perfect for those interested in illicit love affairs: knowing a man was underground for a significant block of time would give another man plenty of time to pursue the miner's wife. But, on the Copperbelt, *iminyama* could complicate matters.

Dalliance and infidelity were long understood as common features of mining township life and the euphemistic term for an affair in some communities had come to be known as "private sector."[7] So, for example, a woman seeing a man other than her husband was said to be having "a private sector." Still, certain rules pertained as to how such affairs were conducted since the wife's behavior on surface could affect what happened to her husband underground. Liaisons were best not done while the cuckolded husband was underground since the wife's infidelity might bring on *iminyama*, subjecting her husband to a rock fall or loss of limb in the mine's machinery.[8]

Iminyama or *chisomo* were also frequently judged to be the reasons why particularly bizarre accidents might occur. One fatality that stumped mine management and workers concerned a miner who pulled supporting timbers down upon himself. Instead of following standard procedure, when given the order as to which supports to remove the miner apparently turned in a different direction and took down other supporting timbers obviously still needed where they were. According to Zambians, the only reasonable explanation for such an event was *chisomo* or *iminyama;* since the consequences of doing anything else were so obvious. Talking about these incidents in the corporate workplace was understandably difficult for miners, however. How could they explain deaths possibly triggered by *chisomo* or *iminyama* when a worker's own mates might not really know what had gone on and when the concept of evil spirits just did not fit with industrial production talk?

Preventive Measures

The companies mounted protection and safety programs in which workers participated. But many Zambians appeared to have additional or alternate means of keeping themselves safe and for identifying the sources of faults when accidents did occur. While some such means appeared specific to the mining context, others seemed to have more versatile applications. Women walking through tall grasses near major Copperbelt towns might tie straw-like strands around their necks and wrists for protection against snakes. Township healers sold shells, bits of wood, and seeds that men could wrap in paper and put in their pockets or tie with a handkerchief around their genitals. These items were said to guard against all sorts of misfortunes in offices, court cases, and so on. Necklaces and forehead tattoos were also available for miners at small medicinal shops in places such as the township of Chimwemwe outside Kitwe.[9]

Beyond employing physical items, members of the workforce understood that use or non-use of certain words could have an effect on miners' safety. One miner's wife, for example, said that each time her husband left for work they had to say good-bye in a special way because they never knew if he would return. Likewise, when he arrived back home they had a special greeting. Even though the words essentially meant, "How did work go?" they carried a connotation that both the man's vigilance underground and the wife's conduct on surface had played a part in his ability to return safely. Miners also knew never to say, "*Mwabombenei?*" ("How is work going?") until after arriving back on surface. Underground, one never knew what might happen.

Another protective measure that some miners employed involved practices traditionally associated with Western Christianity. Zambia is distinct in the world as being the only officially declared "Christian nation." In the mid-2000s, roughly 86% of the population identified as Christian and public conversations frequently contained references or allusions to a shared Christianity.[10] So, quite a number of miners would say simply that they survived underground by the grace of God or that God alone protected them. One man once related that, upon being given a book of Christian prayers, he began copying down a sample each day and placing the bits of paper in his pocket before going on shift.

Miners complemented such preventive and protective schemes with an understanding that supernatural forces or spirits could send warnings and assign blame for what they considered improper conduct toward their region—in this case the underground workplace. These spirits were especially adept at discerning which relationships with humanity were appropriate and which were not, as a newly hired mine physician on an underground tour quickly discovered.

> You know, we try to arrange underground visits for our doctors here so they can appreciate the environment in which the miners are working. And the chap who was leading them around underground, you know, you get ground mov-

> ing—it starts creaking. And he just said, "The spirits of the mine are not happy. Let's leave this place." . . . [T]hat is the sort of thing, you know, they have respect for the mine. "The spirit of the mine is not happy that we are walking in here. You guys" almost implying that "you're not miners you've got no business being here. So let's get out."[11]

Similarly, some Zambians saw a multiple fatality incident at Konkola's mine site as signifying the mines spirits' disapproval of owner, Anglo American Corporation. In 2001, an earth wall collapsed at the Chingola open pit killing 10 miners and stopping production for a month. It was not long after this incident that AAC, reporting huge losses and lack of sufficient funds to proceed with the underground Konkola Deeps site, announced its departure as controlling partner in the company. Popular understanding of the Chingola incident, though, said the mine spirits had already made their decision. Anglo American would have to go; and so they went.

Zambian "Midrash"

This linking of what, to outside observers, might appear to be unrelated factors has precedent and deserves attention for a couple of significant reasons. The first has to do with what it implies about how societies determine relationships between ideas, events, time, and space. In Zambia, an association of events or ideas that deviated from what a dominant development narrative would suggest seemed to occur not just in the case of accident prevention but with post-incident investigation as well. When something did go wrong in the workplace, the reasons Zambians could give for this occurrence would likely not be the same as those mentioned in corporate policy papers.

In fact, the Zambian workforce and community appeared to employ a two-tiered approach when investigating the causes of mine accidents. An outward, or public, discussion would first identify accident causes that lay with the company. The company should have implemented better protective measures, for instance, or production pressures and haste had led to the incident.

At this level of inquiry company managers often experienced laborers' resistance to official fact-finding endeavors. To the workforce, these formal proceedings could be very intimidating. Everyone lined up in the boss's office with summary dismissals seeming always to follow were not encouraging means of producing free discussion. So, from the workers' vantage point, prevarication was not meant to frustrate the company so much as to protect the miner's job.

> [W]ith the type of information gathering that, sort of like, puts the spotlight on the individual they will just lie. They will just lie. Because he says, "I'm saving my skin. I'm saving my job." [If management asks,] "Did you see this? What happened?" "It just happened." "But the way the body was found and so on, it's like something dropped." . . . He says, "No. It's just loose rocks from somewhere." You know? And they would lie just like that.[12]

Beneath this rather obvious level of investigation and blame assignment, a second, more hidden level of inquiry might take place. Focusing on the individuals involved, such exploration would try to determine why the accident happened to a particular person at a particular time. This more private, personal search could take place simultaneously with the company's own formal investigation. Retrospectively, Zambians would try to determine why the incident had happened by examining signs and actions even if they did not appear immediately relevant to the accident.[13] So, family members might, for instance, ask if something unusual had taken place just before the miner left for work that day. Had he, for example, done anything untoward such as showing disrespect for an elderly person? Ultimate conclusions might not be definitive and could encompass a mixture of supernatural and human-derived causes. But they appeared to account, at least in part, for why the misfortune had occurred.

This type of pieced-together inquiry was also not limited to the mining environment. At times, the search for a death's reason could seem astonishingly creative and, at others, poignantly paranoid. On the creative side for example, a Copperbelt resident recalled a young secondary school student who was found keeping company with a female at a time and place that his school headmaster considered inappropriate. He received a reprimand. Then some papers in an office went missing and about that same time there was a funeral. People from the area who knew about all this added the principal's reprimand plus the missing papers together to conclude that they foretold the funeral. More worryingly, in 2004 an Anglican evangelistic crusade took place on an open space just outside Kitwe along the Chingola road. The busy stretch of road from the Buchi township turnoff near Kitwe to Chimwemwe farther out toward Chingola was regarded as particularly treacherous territory being haunted by the many spirits of those who had died there. The day before the crusade began, there was a major accident in that area and a number of people were killed. The next day, at the very same spot a pedestrian was run over. Some Zambians concluded these incidents meant that those running the crusade were seeking people's blood to help ensure a successful event. Likewise, in 2005 a young man who was doing piecework at an educational institution outside Kitwe learned that his grandfather had killed himself. But suicide was unheard of among Zambia's elderly. Furthermore, the grandfather had just come from the village and had successfully collected his son's terminal benefits from the mining company so he would have had no reason to despair over his economic well-being. Yes, he had quarreled with his wife before leaving home, but someone would not end his life "over a woman." With cultural, financial, and relational reasons for the death ruled out, the only reasonable explanation was supernatural: *chisomo* or *iminyama*.

The second reason that this method of inquiry deserves to be taken quite seriously is the means by which it represents a legitimate reasoning process. While the steps of and elements included in that process might not accord with Western cultural expectations, still a distinctly rational thought process appropriate to the cosmological context was present.[14] Further, it could not be assumed that exposure to Western thought patterns would change how Zambians understood things; in fact sometimes just the opposite occurred. My research assistant,

Enock, who was considerably better educated than the majority of his peers and a marvelously analytical thinker claimed to come to a belief in the supernatural snake guardian not because he had grown up in a traditional village but *after* he had gone underground in an industrial copper mine for the first time. Previously he had thought stories of the underground snake and the roan antelope pointing out copper in Luanshya were just African tales. Now he concluded that copper discovery and its mining could only be done with supernatural help.

Enock's change of mind actually represented a compatibility with scientific-rationalist thinking that was evident many times during data collection for this study. His belief did not change his appreciation for the industrial mine's mechanization or his own use of technology. It simply said that science had help from something transcendent. Likewise, the story of William Collier's romanticized hunting expedition lost nothing through popular Zambian interpretation of events. If anything, the Zambian view added a sense of cosmic blessing as the felled roan antelope showed Collier precisely where to look for copper. One further example bears some extended mention in this regard. The chief's nephew quoted in chapter four with reference to underground snakes also recounted an incident about discerning messages from the spirits at the Kansanshi mine.

> Elo, ifi fine ifya mipashi fyena fyacishinka naba abaleisula maine kuSolwezi, balibulile amadrinks yabili, Coke na Fanta, baingisha pa underground pamaini, kuKansanshi, baingisha amadrinks yabili elyo baletampafye pa surface pamulu. So after babika filya bapepelela no kupepelela filya fine bafipepelela, baishile sanga ukutila Coke basenda, pashala Fanta. Again baisa bula naimbi Coke, naimbi Coke na Fanta pakati, so icalesanda yalya amadrinks cindo? Yali mipashi, ilya Coca Cola yale-mininga ukwebati nifwebo fweba nkande ifitile, fwebene bacalo, fwebali koti kuno kwa baNkana. Elyo Fanta niaba abanensu abeshile mukutwafwako imibombele, ifyapala abasungu eba Fanta, so baishile mukusanga futi pafuma, Coca Cola na Fanta inga itatu. Nomba ninshi paya amabotolo yanga yatatu. So yabili ya Coca Cola imo ni Fanta. Ifya lesendefyo mipashi ne kusuminisha ukwebati okey kanshi abafitile koti nifwebo fwebene babukuno kwesu ne bakashikile na abasungu kuti twabombela pamo, nomba fwe bengafulamo nifwebo fwebafita.

> About the Spirits, these things are real. For instance, those mining at Solwezi took two soft drinks, Coke and Fanta underground when they were just starting on the surface. So after placing them underground, they prayed to the Spirits according to their traditions. They later found that Coke was taken and Fanta was left. And again they took Coke, and another Coke and half of Fanta. So what took those drinks? The Spirits! Coca-Cola meant it is "us" the blacks like the people of Nkana, who are the owners of [the] land. Fanta refers to our friends, who have come to help us in our work such as the whites. They found Coca-Cola was taken and only Fanta was left. This after three bottles had been taken. Two Cokes and one was Fanta. The Spirits used to carry these drinks in order to confirm that, ok, the black people are the owners here, however we can work with the whites but we must be in the majority.

Distinctive empirical elements are present in this trial conducted by the miners, for note that they tried the Coke-Fanta experiment twice by way of testing their findings. Further, to the miners and the nephew who recounted the story, evidence produced by this experiment was sufficient to offer proof of the mine spirits' existence as well as socio-economic commentary on how the mines should be run. That soft drinks were used in ways unintended by their own producers was, within this setting, immaterial.

Zambian Narrative at Work

Cultural tidbits such as these may be of interest to outside observers and those involved in professional policy making. But the odd anecdote usually fails to gain sustained interest in the work place, typically losing out to the crush of other concerns.

There is, however, a larger philosophical issue present in such stories that deserves sustained attention. This concerns the underlying alternate narratives that such anecdotes may represent. As we observed at the outset of this study, how individuals and groups piece together the ideas and events of their lives plays a huge part in determining how they act. So what did the various incidents as recalled above signify regarding how Zambians interpreted what was happening with their lives? How did the stories Zambians told themselves affect their understanding of the privatization process?

A really bad series of misfortunes suffered during the course of this investigation offered some clues. But since this series of incidents must be set against the backdrop of an earlier story, we need to revisit briefly a bit of Zambian mining history before retelling the events of 2005.

The Mufulira Disaster, 1970

Not long after the Kaunda government announced its intention to assume a 51% share in the mines, the worst industrial accident in the country's history took place in Mufulira. Mufulira is a town near the Democratic Republic of Congo border, which at that point draws a somewhat arbitrary line across the road leading toward Zambia's Luapula Province.[15] Without the mine there would probably be no reason for the town and, at first glance, Mufulira seems more isolated than the Copperbelt towns of Kitwe or Ndola. Internet access, for example, is sparse by comparison and there are not as many shops and restaurants as can be found elsewhere. But, with a closer look, Mufulira may have always been less isolated than it appeared in the mid-2000s. It has a long-standing role as a way station for those traveling south and west from Luapula through the Congo, areas where supernatural activities are said to be particularly strong.

The night of September 25th, 1970 a tailings dam on the north side of the waste rock dump collapsed and more than one million tons of mud and water slid into the mine. In fifteen minutes, huge sections of the mine were inundated and only thirty-one miners on duty there emerged alive. Another 89 mostly Zambian workers were counted as dead, the bodies of at least 50 never being recovered. In addition to this human tragedy, the Disaster's immediate effect on Zambia's copper production was monumental. Monthly production dropped to about a fourth of what it had been right before the sinkhole collapsed.[16]

Official reports from the time of the Disaster up until the present day attributed the incident to a variety of mechanistic causes. The sinkhole was an old problem, warning signs for which had been missed, ignored, or perhaps even covered up for decades.[17] Something simply happened that September night to make the old mine tailings more liquid than solid. Scientifically speaking, this could occur under the right circumstances—an earth tremor for example.

But such an explanation was insufficient for the general Zambian populace. At the time of the Disaster, rumors circulating about the actual cause of the accident included supernatural forces and malevolent impulses.[18] For decades afterward, many Zambians suspected the cave-in had been planned. One miner who had grown up in Mufulira was a teenager at the time of the Disaster and, in his experience, published reports concerning the numbers of dead were woeful underestimates. He recalled at least fifty funerals in his own township section while others had thirty to one hundred funerals. Looking back, he said, the real reason the Disaster had occurred was because whites, upset over nationalization and Zambianization, secretly dug tunnels in the direction of the slag dump to provoke a cave-in. Resentful that they were being chased from the mines, as Robert Mugabe would later chase whites from Zimbabwe's farmlands, expatriates committed sabotage.

Decades later, even though the underground area was sealed off, the old tailings dump no longer used, and mine development was proceeding in an opposite direction from the inundation, the Disaster continued to affect the Mufulira community. Some present day workers there said that occasionally body parts were still being retrieved. And an old European cemetery in a lone grove of trees near the mine was understood as the burial ground for those whose had died in 1970. Even though the Disaster's specifics were fading from general public memory there still seemed to be something significant about the place where it had happened.

> As we were growing up I remember the area which—where the tragedy happened—I don't know the way things happen. There's a lot of guava trees; a lot of guava trees. So it has been put to be a restricted area, you know? It has been put as a restricted area one, on the basis that the place is not safe to go there. You can easily fall into a ditch because there are still some openings. Once you fall into a ditch you may not be able to come out. . . . The other thing is there is a belief to say the people who died from the same tragedy, you never know if you go there maybe you can—you may not come back! You will not know what has taken you. So it's somehow restricted. Though I'm sure those were beliefs of those during those days when memories were still fresh. As it is now, well, people are, they are tending to forget about it. I remember while I was in Mufulira as we were going to our fields we could pass through the same place. It's no longer that restricted now. It was really a tragedy.

As to the guava trees,

> Well, this is what surprises me. I never really inquired as to how the guava trees found themselves to be that side. I don't know whether they were just being planted—of which I don't believe they were just planted. I don't know. All I know is there is a lot of guava trees.[19]

The Mufulira Disaster was not something of official concern to managers in the privatized mining companies in the mid-2000s, the incident represented historically useful information but was no longer relevant otherwise.

In casual conversations with everyday Zambians, however, the Mufulira Disaster would spontaneously arise as a topic. When it was discussed in such settings, there could be an uneasiness to the memories suggesting that why it had happened and what it might mean were still unanswered questions.

> If you were to go to Mufulira and try to mess about with the area where there was the Disaster, you would be in trouble with the miners. . . . You have to have respect for that particular area. That particular incident happened and, over the years, people have forgotten the details—the geological, the mining reasons—why you had the Mufulira mine disaster. But everything that is related to that, you know, or the area—you don't go messing about—it is held with the same reverence that they do, you know, or that we had for graveyard cemeteries. You don't go messing about there . . . you don't point at the graveyard and you never went through the graveyard. In fact, in those days, people didn't die that often. But [with] AIDS now and poverty and so on that has changed. But, here—the mine itself—that disaster itself some would say the mine was not happy with something so it happened.[20]

With this background in mind, what happened in early 2005 could be seen as genuinely indicative of affairs gone awry.

The Spate of Accidents, 2005

Five years after the last of Zambia's mines had been sold, a series of accidents prompted heated debate about the country's privatization program. The precipitating event was a multiple fatality incident at Mindola Sub-vertical shaft in February. Miners working at the site came on foot from townships nearby such as Mindolo North or Chimwemwe, or were bussed in from Wusakili further a field. Shortly after this incident, near the entrance gates to the plant site a crowd of miners' wives gathered, demanding dismissal of certain managerial staff. An expatriate shaft manager was temporarily removed by the Mines Safety Department and after a few days all again seemed normal.

Then, in mid-April a truck carrying school children from Kawambwa, Northern Province overturned and 45 young people were killed. Accidents of this type earn news coverage in large countries such as the United States. For a small nation such as Zambia, the event warranted headlines for days.

Things got worse less than three weeks later when, back on the Copperbelt, a huge blast destroyed the BGRIMM Explosives factory. The factory, owned by

Chinese investors, sat on the grounds of Nonferrous Metal Mining Corporation (NFC) Africa Mining near Chambishi. The story of what Joseph's family experienced that day was probably typical of what other families encountered. The morning of the blast, father, mother, and their twenty-one year old son Joseph had gotten up early to go for morning prayers at their church. Joseph and his father had then gone to work together, Joseph at BGRIMM, his father in the rewinding department of NFC. When the explosion occurred, Joseph's father heard it but could not determine the direction of the blast. When it became apparent that the explosives factory was involved, he tried first calling Joseph's cell phone and then text messaging him. But he got no response. He then telephoned his wife and went to the plant site. There, rather than seeing the large steel supported building where his son worked, the father found only rubble. Later in the day he found his son's body, recognizable by the clothes he had been wearing that morning. Once he found his son, Joseph's father tied a cloth around the young man's ankle to guard against mix-ups when time came for official identification.

The next few days were chaotic. According to the first media broadcasts from the site, families that gathered to locate and mourn their dead shouted such questions as "Is this the price we have to pay for privatization?" into the cameras. President Levy Mwanawasa,[21] who received news of the explosion as he was about to board a flight from Ndola departed for Lusaka rather than hastening the few miles back to Chambishi. When he eventually arrived, he was initially upstaged by former President Frederick Chiluba who had gotten there first. Both men met with grieving families at a nearby schoolhouse to which they had been shunted. But to mourners, the President's delayed appearance and the haphazard arrangements for gathering together suggested that government was not taking their suffering seriously.

There were harsher words for the Chinese investors and managers, however. According to some reports, local Chinese supervisors had told workers and some eight apprentices from Copperbelt University (CBU) that temperatures in the plant were too high. They asked everyone to address the problem and left. Fifteen minutes later the building exploded. To observers the conclusion was obvious: the managers had been aware of the problem and gotten out in time to save only themselves. The eyewitness who survived to relate this story was a young man who reportedly had been blown into a barrel that subsequently rolled to safety. Now speculation was that the young man's life might be in danger for he knew more than management would want disclosed.

Public antagonism increased when company officials' first statement in the media contained a promise of financial compensation for victims' families. Recovery and identification of victims' bodies were ongoing so it seemed to Zambians callously out of place to talk about money right then.

In truth, ordinary Zambians never really expected the recovery to be complete. Most of BGRIMM's employees had been poorly paid casual workers and personnel record keeping was highly suspect. When authorized reports came out with estimates of between forty-nine and fifty-one people killed, popular opinion suggested a significantly higher total. In fact, rumors went around that a number of the coffins eventually assembled actually contained only body parts and stones. Zambians said the owners knew the blast had taken many more lives than they could account for. This explained why their officials hastened to Kitwe Central Hospital to witness the caskets being filled and sealed. Zambians also understood why more people had died than the media reported. After all, just a small amount of explosives could pulverize rock. Imagine what would happen to human bodies torn apart by the blast of an entire factory: the people inside would have been vaporized. So any found body parts were more likely from passersby near the factory gates than from BGRIMM workers.

People were particularly outraged that the Chinese managers had apparently sacrificed Zambians while saving themselves. If this were truly an accident, why had no Chinese employees died? The day of the funerals some CBU students faced off with President Mwanawasa's entourage. They said they wanted Chinese buried along with their fellow citizens. After police handcuffed the student protestors, however, mourners of the dead appealed for their release. They said proper grieving was impossible when others were in pain.

Subsequent to the explosion, Joseph's father said his wife had also been ill and in hospital. He himself had suffered a bout of malaria and a large boil had appeared on his buttocks. Their daughter had been ill as well and his brother, a miner at another company, had gone missing.

Public speculation concerning why the disaster had occurred began immediately after the explosion. Newspaper commentaries blamed the privatization process for bringing in inexperienced workers. They also clearly suggested that this sort of thing was related to Zambians' past experiences. The Mufulira Disaster, sporadic ownership of the Luanshya mine under privatization, and the "unprecedented spate of underground mine accidents"[22] beginning at Mindola were ills that resulted from improper observance of safety procedures. Similar talk excoriating the privatized companies for failing in their responsibilities dominated public discussions and the media over the next several weeks.

In private, or casual, conversation speculation about reasons for the incidents encompassed different subjects and suspicions. The day of the BGRIMM blast, a number of local people began accusing President Levy Mwanawasa of sacrificing Zambian citizens to Satan. The numbers of dead in Kawambwa and Chambishi also seemed significant and linked to greater instances of carnage. Then there were memories of the fateful April in which Zambia had lost its soccer team in an air crash. Why had both tragedies happened in April?

The series of accidents extended into May. Nine people died in a car crash near Kabwe. Rock falls at Nchanga in Chingola and at Luanshya killed one miner each and the president of the Mine Workers Union of Zambia (MUZ) wanted mine owners charged with murder for such on-the-job fatalities.[23]

Then, on May 18th Mufulira was hit again. A cage door in one of the shafts had come open and some miners had fallen to their deaths. Roger, a casual worker with the mines, watched events unfold as he waited to find out about his uncle who had been on shift that day. By afternoon when he met with friends at the bus stop, he still had no information. So, they all decided to go to a local pub and wait for more to emerge. On the way they could see and feel tension among people moving about in the township. Someone shouted in Bemba to an anonymous white woman driving by: "You are the ones who are killing our people!"

At the entrance to Joggers Pub, two or three men stood around animatedly recounting what had gone on a few hours earlier. "These guys are killing us," said one about the company. Another cluster of men talked out back in much the same terms. But they also speculated about witchcraft and possible Satanic sacrifices; the sequence of events being just too odd not to suspect the extraordinary. In addition to everything else that had been happening in the country, barely forty-eight hours before this latest incident the mining company had delivered about 12,000 plastic bags to its hospital in Mufulira. Management was said to want them for body collection should something like the Chambishi explosion take place. And now this. How could management have predicted it? Something must be going on.[24]

Back in town, family members and friends confronted police at the mine site and at the hospital where the injured were being treated. ZNBC-TV recorded an angry encounter between government and mining officials and some people threw rocks at passing vehicles. Police responded with tear gas.

By the end of the day official reports held that two miners had died and four others were missing. Popular assessment was of a much higher death toll and, as news of the accident worked its way around the Internet, estimated fatality numbers varied from the fifties to the hundreds. What exactly had happened was something of a mystery because mine cage doors open inward. In this case the front door of the three-tiered apparatus—large enough to hold half a 747 jetliner—had come open, perhaps caught on something. A vacuum must then have ripped off the back door as well. Workers nearby had fallen out, perhaps grabbing onto others as they fell.

That same afternoon a Zambian military airplane crashed in Mongu, Western Province killing all thirteen people aboard.

A couple of days after these two accidents, ZNBC-TV News produced a special report featuring Minister of Mines Kaunda Lembalemba, a former miner and an ordained Pentecostal preacher. In the early 2000s Lembalemba had had a regular evangelistic television show of his own. Asked how he accounted for the spate of accidents, he offered that they were attributable to spiritual warfare.[25]

Other public officials and private citizens also speculated that events bringing death to some Zambians were indicative of larger struggles. On radio, Michael Sata, the leader of a prominent opposition political party, declared that things such as the BGRIMM explosion were deliberate attempts by investors to reduce the Zambian population.[26] Popular speculation also continued that President Mwanawasa, having run out of family members to sacrifice in his quest for sustained wealth and power, had begun offering up miners.

Man: Mwalishiba ifintu ngafi lecitika ifisoswa fingi, palasoswa ifintu ifingi sana. Nabaleti ati niuyu wine President (kabili mulefwaya tulande ifilecita) ifi filefuma uku kwina bekala	Man: You know when things are happening, much is said. Some people are saying it is this very president (didn't you say you want us to say what is happening?). These things are being

kuLusaka uku. Ababene (ba president) eko baile kuIndiako mukupalwa, kuti aya shani mukupalwa kulya kuli mwenye. Ukucila kumuntu munabo. Ukufuma apo babwelela kulya ifintu fyalikonkana: banyina bafwa, ama nieces na brothers bafwa.

caused by the president, who lives in Lusaka. The president went to India to be blessed. In any case, why should the president seek blessings from Indians rather than from his fellow black people? Since he came back from India, disasters have been happening one after the other: his mother died, his nieces and brothers died.

Another man: Ico bailiile kuIndia kukupalwa, babapela nama rice ayengi sana. . . .

Another man: When he went to India to be blessed, he was also given plenty of rice. . . .

Man: So ukufumapo ifintu nafikonkana sana ukufuma apo bafumina musamba kuIndia. Abantu baleilishanya daily. . . .

Man: Since he returned from having been cleansed[27] in India, bad things have happened, one after the other. People are complaining daily. . . .

Another man: Ifyo mwalanda bamayo cinshika banyina waba president ukufwa muli minibus! . . .

Another man: What you have said my mother is very true, the mother of the President dying in a mininbus! . . .

Interviewer: Kwaliba imfwa shimo shilya balanda ati bamuposelekofye, kuti mwa landapo shani?

Interviewer: There are some deaths that they say s/he was just thrown him/her there, what would you say about it?

(Response): Eifyo tulandapo ati babwelafye kuIndia—kuti twayeba ati chisomo.

(Response): That is what we're saying—that he just came back from India—we can say *chisomo.*

Interviewer: Kuti mwa citondolola shani (ichisomo) kuli ine ne ushili umubemba?

Interviewer: How can you explain (*ichisomo*) to me who is not a Bemba?

Woman: Kuti twalondolola ukutila balya baile ku kupalwa kuIndia, after fye balebwela kuIndia, bapalilwa muli ba nyina bapya umulilo ati bapya banyina. Babalilapo umulopa/umubili waba nyina ba ucita sacrifice, after ba cita sacrifice umubili wabanyina nomba baya ukulakonkanyapofye abantu babo, na bana bene aba kumasukulu uko bafwa, bene nimpiya shonse shilya nimpiya shalitulama muState House. Uku bashimine balefwafye.

Woman: We can explain like this: that man went to be blessed in India, just after he returned from India, he was "blessed" through the death of his mother in fire. He started by sacrificing his mother's blood/body, then his own relatives, then school children died. What he is after is money which is piling up in State House, while the miners continue to die.

Another woman: Talangako (president) nobulanda filya ati alemonekako no bulanda awe. Chiluba alelilako mwamonako nefilamba. Nomba uyu ena awee. Aumfwa na accident anina aya. Elyo nama church yafula. Mailo apepako uku aleka apepako uku ale-

Another woman: He doesn't even show sympathy, not at all. Chiluba would cry and we saw his tears. But this president does not. When he hears of the accident, off he flies away. In addition, he has claimed membership of many churches. The other day he

ka—president, icimweneno cesu cilesankanyafye ama church. Umuntu asha abantu bali muaccident, anina indeke aya asha abantu balelosha.	worships in this church; he stops. Then another church; he stops—President is our role model but he is the one who mixes[28] churches. He flies out of the country, when people are in an accident and families are moaning.[29]

The mining incidents, although statistically claiming fewer lives than the roadway and air tragedies, ultimately garnered the most public attention and heated debate.[30] No accounts exist of the public protesting in support of road safety on a level matching the demonstrations and debates that occurred regarding mine safety at the time.[31] But so powerfully did the entire series of events grasp the national consciousness that, on Sunday May 23rd, an emergency prayer service took place at the Anglican Cathedral of the Holy Cross in Lusaka, Zambia's official house of prayer. Organized by the three national Christian clerical bodies, the service was promoted as an opportunity for Zambians to pray concerning why the series of recent accidents was occurring and to commemorate those who had died. The homilist that day wondered,

> "What is God telling us? . . . Let's together plan for the safety of our people, even the unborn. All these accidents, not only in the mines, but also road traffic accidents, could mean that maybe there is something that we have done."[32]

President Mwanawasa attended and read from Psalm 46.[33] Former Presidents Kaunda and Chiluba sent condolences and urged the Zambian people to pray.[34]

A few more miners died at Nkana during June and Mines Minister Lembalemba threatened to withdraw the private company's operating license.[35] Company officials shot back with a public challenge to this threat and the rash of accidents then seemed to subside.[36]

But popular argument over the incidents continued for some time with accusations and blame being spread widely. The government and miners' unions prevaricated about who was responsible for what with their only point of agreement being that the mining companies valued production over worker welfare.

> Lembalemba said while acknowledging that most accidents did not just happen but were caused, some of the accidents occurred due to negligence by the mine owners while others were caused by the mindset of workers themselves.[37]

> Accidents just do not happen. They are caused. The cage is one of the most important machines . . . and maintenance is often overlooked. . . . These deaths are artificial.[38]

> Mr. Mutale said with the current situation, mine owners were always blaming the dead miners because they were in the forefront of making a report on how the accident happened.[39]

The media and ordinary Zambians also generally agreed that the investors' ruthless pursuit of quick profitability had come at the expense of safety measures.[40] In fact, such accidents were simply evidence that the companies barely regarded miners as human beings.

The general public often claimed that nothing like this spate of accidents had occurred when the mines were nationalized but they did not completely exonerate the government from its share of the blame. Generally, there was disgust at the way Zambia's government had implemented the privatization program; disgust at the President's initial non-responsiveness to the BGRIMM explosion; and disgust regarding how various officials were thought to be advancing their own economic positions while ordinary citizens suffered.

Interestingly absent from debates about the spate of accidents was analysis of individual Zambians' potential ability to have pre-empted problems when, on at least some occasions, warning signs and opportunities for people to take preventive measures had been present. Zambian police, for example, had allowed the obviously overloaded Kawambwa truck to pass a roadblock in exchange for a ZMK60,000 bribe (about $15 USD at the time) and the driver apparently knew he had faulty brakes. A young female BGRIMM worker was recalled as relating misgivings to her boyfriend about the plant's operation. But neither she nor he acted ahead of time.[41]

By late June, the promised government investigations into the spate of accidents had not progressed very far and victims' families remained frustrated.[42] Irritation expressed by government and unions toward the mining companies was generally reciprocated. And, as to the investors, there was largely silence.

In late July, mining explosives were discovered on the roof of a house in the isolated township of Mpatamatu near Luanshya and miners at KCM embarked on a violent strike over disputes during union salary negotiations.[43] Michael Sata, the politician who had speculated about investor attempts to reduce Zambia's surplus population, claimed responsibility for inciting the riots and President Mwanawasa ordered his arrest on sedition charges. By early August, Minister of Mines Lembalemba had been removed from office as well.

Points of Relevance and Wonder

Several interpretive points from these stories deserve mention. The first is the sense in which both the Mufulira Disaster and the 2005 spate of accidents elicited powerful life and death fears on multiple levels beyond the physical. These encompassed the economic and spiritual as well. From a development policy making standpoint, worries about white sabotage and talk of efforts to reduce the Zambian population would likely demonstrate sheer paranoia of the people or opportunism on the part of politicians interested in escaping their own complicity in the problem. But why would citizens of a nation-state that was, by then, more than two generations old think this way and how would opportunistic politicians know that such ideas might effectively manipulate the citizenry?

A second point of note concerns the prevalence of connecting what to Western policy making minds may appear as unconnected ideas and events in order to draw conclusions about why the tragedies happened. In the Mufulira

Disaster story, popular views associated Kaunda's mine nationalization program with racial tensions and those two things together formed at least a partial basis for why the 1970 collapse had happened. During the 2005 spate of accidents, the Mufulira Disaster itself was resurrected. It, along with the air crash in which Zambia lost its football team[44] and accidents in seemingly unrelated settings throughout the country appeared indicative of a supernatural, or cosmic, judgment rendered on Zambia's overall conduct.

Before we dismiss this chain of reasoning as indicative of undue suspicion, poor education, or political opportunism we should ponder a third note of consequence. This is that the complex relationships and multiple motivations extant among the various actors in this story defied dualistic or binary divisions between management and labor; black and white; expatriate and national. That being the case, how well could managerial theories, economic, political, or historical analyses used in isolation adequately deal with such a variety of tensions and sympathies? In other words, attributing Zambian perspectives to typical frictions between workers and bosses or between different races and classes would certainly have merit. But it would also seem to produce an inadequate analysis. Therefore, when standard development policy and planning discussions are underway in situations of radical cultural difference it is crucial to be cognizant of their conceptual limitations. In this specific case, for example, where should we fit such factors as the brutal analyses of ordinary Zambians concerning their own government leaders' spiritual activities? What of the passionate protectiveness the people had for the mines themselves? What, further, can be made of the government's stated aim of restoring the mines to pre-eminence followed by its antagonistic actions toward the workforce and managers?

Clearly, Zambians drew upon a variety of modes of thinking when assessing their individual and collective well-being. And analysis that relied upon religious or spiritual interpretations simply could not be divorced from this mix. Moreover, it appeared that these concerns were not merely some elements among many but were quite central both to *why* things seemed to have gone wrong and to *how* people inoculated themselves against further calamity. The next chapter will explore this possibility in greater depth.

Notes

1. This may also have been an indirect reference to the long-standing practice of selling uniforms. White observes that clothing or uniforms issued by the colonial era mining companies were considered commodities by the African workforce. New issues of gear were often fairly quickly converted to cash. *Speaking with Vampires*, 290. Consequently, contemporary era complaints about shoddy gear could also have indicated the miners' loss of a small, but crucial income source.

2. Even though their task was specifically to study African Christian expressions on the Copperbelt, Taylor and Lehmann observe that Christianity appeared not to alter fundamentally African worldviews with regard to the importance of maintaining good relationships with all of creation. In that view, well-being is derived from harmony with na-

ture, humanity, and the unseen. When things go wrong, a person is assumed somehow to be out of step with this order. *Christians of the Copperbelt*, 275-76.

3. The extent to which younger generations knew about this traditional concept was questionable due to the increasing pervasiveness of Western influences and continued break up of family structures.

4. Physician, Copperbelt mining company, interview with author, 2004. Taylor and Lehmann report that, in so far as they could discern, the Bantu worldview contained "no distinction between a natural and a supernatural order of being." *Christians of the Copperbelt*, 275. See Ellis and Ter Haar on why these sorts of discussions happen in African everyday life even if such talk, in the West, is typically left to theologians and philosophers. *Worlds of Power*, 26.

5. Similar instances were known to occur in South Africa and Mozambique.

6. Physician, Copperbelt mining company, interview with author, 2004.

7. See similar references in Epstein, *Scenes from African Urban Life*; Ferguson, *Expectations of Modernity*; and Powdermaker, *Copper Town*. Additionally, Parpart's Copperbelt study investigates the complex reasons for infidelity and especially the resilience that African women demonstrated in working to achieve some measure of economic self-sufficiency. At the time under Parpart's consideration, a common slang term for affairs was "playing." Parpart, "Class and Gender on the Copperbelt," 15-17, 20, 22.

8. The point here is not so much to comment on what this might say about gender and power relations but about transcendence of time and space.

9. The conflicted regard for such practices within the contemporary, urban setting meant that Zambians who sought these services sometimes went to townships other than their own out of concern that their efforts would be witnessed by neighbors otherwise.

10. According to 2005 estimates, 9,605,919 out of a population of 11,478,317 identified as Christians. http://www.worldchristiandatabase.org/ezproxy.bu.edu/wcd/ (accessed April 1, 2009).

11. Physician, Copperbelt mining company, interview with author, 2004.

12. Physician, Copperbelt mining company, interview with author, 2004.

13. The reference to Midrash calls to mind the issue of interpretation. Jacob Neusner, for example, maintains that traditional rabbinic Midrash operates on the assumption that everything in Scripture was recorded for a reason. Therefore even ostensibly superfluous passages contain important teaching elements. Further, all Scripture is inter-related and a verse from the prophetic writings may be used to explain something in the Torah and so on. *The Way of Torah: An Introduction to Judaism* (Belmont, CA: Wadsworth Publishing Co., 1993), 54-63. In a similar way, many Zambians appeared to ascribe relational meaning to events that most Westerners would see as superfluous to the issue.

14. See note 76 in chapter one. In commenting on the difficulties that adherents of a modern, scientific-rationalist worldview have in accepting any other forms of logic and reasoning. Fields writes, "To us, a false prediction reached by a recognized method is not comparable with a true prediction reached by faith. Faith promises us nothing certain. Correct method promises us salvation form error—certainly, if not immediately." *Revival and Rebellion*, 272.

15. See White on the comparative ease with which Africans historically crossed back and forth between the Northern Rhodesian and Belgian Congo borders. *Speaking with Vampires*, 275-76.

16. See note 127 in chapter three for statistics.

17. "Danger That Went Unseen: Muf Inquiry Reveals Wrong Diagnosis," *Zambia Daily Mail*, February 18, 1971. Reports dating from the 1970s and available at the ZCCM Archives talk of studies that, in retrospect, should have been made 20 years previously.

One independent technical consultant also recalled credible reports of mud oozing in two days before the Disaster. Apparently the expatriate manager who had seen it was "banished to Nampundwe." Interview with author, 2005. It is important to remember that, in 1970, the expatriate management presence in the mining companies would still have been substantial. So, blame assigned by contemporary executives of the mid-2000s to management of that earlier era implicated expatriates as well as Zambians.

18. A number of miners also appeared to object that the published versions of the commission's inquiry were too complicated and technical for the layperson to understand. "Miners Want Muf Report Translated," *Zambia Daily Mail*, February 17, 1971.

19. Former Mufulira resident and mining family member, interview with author, 2004. This remark may be an allusion to, or at least historically and culturally informed by, the tradition of revering natural shrines such as groves of trees. See Binsbergen, "Explorations in the History and Sociology," 50.

20. Physician, Copperbelt mining company, interview with author, 2004.

21. Levy Mwanawasa was Zambia's third president. He served as vice-president under Frederick Chiluba and died in 2008.

22. "Sikota Wina on Tuesdays: BGRIMM Explosives," *The Post*, April 26, 2005, 5.

23. "Mine Owners Should be Charged for Murder," *Zambia Daily Mail*, May 16, 2005, 2.

24. A member of the managerial staff subsequently explained that this incident represented a routine supply delivery that coincidentally took place close to the time of the cage crash.

25. "There is this wind of accidents in the country which is very, very bad. And I think us, on the side of the Church, it's important for us to say to you that there must be some wars in the heavens. There must be some wars in the heavens because I have not seen such accidents happening in the country, be it mining or road carnage. I've never seen such accidents happening in the country. I definitely think there must be some war in the spiritual realm." Kaunda Lembalemba, interview on "Special Report," ZNBC-TV, May 20, 2005.

26. This was also during the time that Zambia had achieved the HIPC completion point in relation to its international debt with the World Bank and the IMF. A similar concern about the potential extinction of the entire Zambian populace was overheard on a minibus that Zambia's having met the HIPC Completion Point in 2005 actually meant reaching the People Completion Point.

27. I.e., bathed. See also Ellis and Ter Haar's discussion of Kenneth Kaunda's use of Indian mystics during his presidency. *Worlds of Power*, 70-72.

28. The translator notes that the word used for "mixing" here is the same one employed when speaking of a man freely switching girlfriends.

29. Interview with author and research assistants, 2005.

30. A prominent Copperbelt folklorist and businessman even referred to the BGRIMM incident as Zambia's 9/11.

31. It also bears noting that public demonstrations (when they happened at all) regarding exorbitant death tolls from HIV/AIDS and malaria were calm affairs usually constituting marches or rallies at which T-shirts were distributed. They did not carry the urgent sense of outrage that the mining accidents elicited.

32. "Let's pray at all times—Levy," *Zambia Daily Mail*, May 24, 2005, 1.

33. "Let's pray at all times," 1.

34. "Only God Can Comfort Us—KK," *Zambia Daily Mail*, May 23, 2005, 1; "Levy Urges Zambians to Approach God at All Times," *The Post*, May 23, 2005, 1; "Christians Pray for Peaceful Zambia," *Sunday Times*, May 23, 2005, 1; "Zambia Needs to Turn to

God—Chiluba," *The Post*, May 22, 2005, 1. This latter article also contained pointed references to April as the month when the series of tragedies began. A further day of prayer to seek God's protection was set for June 4th. "Do Not Gain from Disasters," *Zambia Daily Mail*, May 27, 2005, 1.

35. "Zam Threatens to Close Mines," *The Namibian*, June 27, 2005, www.namibian.com.na/2005/June/marketplace/05BE825C8F.html (accessed June 27, 2005).

36. "Mopani Demands Government's Position Over its Fate," *The Post*, July 1, 2005, http://allafrica.com/stories/printable/200507010292.html (accessed July 3, 2005); "Zambia to Let Mopani Mine Operate for Now," *Creamer Media's Engineering News Online*, July 1, 2005, www.creamermedia.net/eng/news/today/?show=69928 (accessed July 3, 2005).

37. "Lembalemba Puts Blame on Mopani Management for Accidents at Nkana," *The Post*, June 27, 2005, http://allafrica.com/stories/printable/200506271215.html (accessed July 3, 2005).

38. MUZ deputy general secretary Rayford Mbulu quoted in "Two Die in Mine Accident," *Zambia Daily Mail*, May 19, 2005, http://www.daily-mail.co.zm/front03.htm (accessed July 3, 2005).

39. "Mine Owners Should be Charged for Murder."

40. "Mufulira Mine Floods Again," *Times of Zambia*, March 29, 2005, 1. In fact, there was widespread doubt that companies were investing at the rate and with the quality they said they were. Many Zambians claimed that old materials were routinely being refurbished and passed off as new. This sentiment was not limited to black Zambians as a white Zambian small mine and contracting company owner expressed the same doubt. General speculation was that the companies showed new equipment purchases on their books but actually just repainted old trucks, tanks, and so on for a fraction of the stated costs. Numerous comments in the newspapers referenced the current mines' poor equipment and expectations that the private companies should be doing something about this. Public coverage and analysis also highlighted the issue of casual labor. "No Zambian Should be Employed as Casual Worker—MUZ," *Times of Zambia*, May 12, 2005, 2; "The Evil of Casualisation," *The Post*, May 22, 2005, 23. From a managerial standpoint hiring casual or contract workers was extremely sensible as it was much cheaper. In Zambia as in other Southern African countries hiring casual workers was also more attractive from a corporate viewpoint because the local laws governing employee taxation and benefits were so onerous. From a practical standpoint, contract workers frequently went to the most dangerous points of operation and were more commonly injured or killed than the mining companies' permanent employees. They also received fewer company benefits and lower salaries.

41. The phenomenon of not taking pre-emptive action seemed apparent in the mines decades earlier to Heron who wrote, "There was fairly general recognition of the fact that *it was not at all the practice for underground miners and workmen on the surface to warn one another of potential danger*, or on their own initiative to take preventive action." *Accidents in the Zambian Mining Industry*, 6 (emphasis in original).

42. "Public Inquest for BGRIMM Victims Fails to Take Off," *The Post*, July 7, 2005, http://allafrica.com/stories/printable/200507070875.html (accessed July 20, 2005); "AF Plane Crash Report Inquiries Concluded," *Times of Zambia*, July 19, 2005, http://allafrica.com/stories/printable/200507190340.html (accessed July 20, 2005). The official report on the series of mining accidents during the main spate of accidents was completed and sent to government late in the year. "State Receives Report on Mine Accidents," *Zambia Daily Mail*, December 6, 2005, 2.

43. Mpatamatu was traditionally a township where miners working at the Luanshya sites lived. Since only miners would have had access to explosives, this incident was further indication of the community's volatile frustration.

44. On April 27, 1993 an airplane carrying virtually the entire national team and coaching crew crashed after a refueling stop in Gabon while enroute to a World Cup qualifying match in Senegal.

♀

Chapter Six

"Jealousy Is There": Accounting for Disparity, Ensuring Success

Development as an ideal often conveys an optimistic view of the world in part because of its close association with economic growth.[1] It is no coincidence that the age of development and an era of unprecedented human prosperity have overlapped. Although parity between rich and poor worlds has been slipping since the Industrial Revolution, the 1980s and 1990s saw a particularly rapid rise in the Global North's ability to attain a measure of economic prosperity over that of those living south of the equator.[2]

By the early 2000s, this inequitable state of affairs seemed somewhat normalized in the parlance of both North American financial institutions and professional development interests. "Wealth management," for example, a phrase abounding in competence and hope, had become part of America's commercial lexicon as if to say that the country's long pursuit of happiness was logically producing well-deserved, rich yields. The greatest societal worry associated with such a prosperous situation was figuring out what to do with it all. For the rest of the world, however, other descriptors applied. In the developing world—especially Africa—"poverty alleviation" was the order of the day. This phrase, especially when overtly linked to the idea of development, contained its own connotations of hope and competence. These connotations could be detected in the Global North where celebrations of the MDGs and related efforts to reduce the number of persons living on less than one dollar per day were popular throughout the first decade of the 2000s.[3] Casually incorporated into everyday talk such phrases subtly suggested both what should rightfully be expected in North America and what could be considered good enough for Africa. A dominant narrative easily reinforced such a dichotomy by maintaining that development was what had secured Americans' prosperity. Surely it could do the same for others.

Those in the Global South, specifically Zambians during the privatization era, seemed to experience a version of this North-South economic dichotomy peculiar to their situation. Public relations campaigns promised that implementing Structural Adjustment Programs such as selling off the mines provided the surest, perhaps only, way of achieving the sort of prosperity they glimpsed through imported media images. But Zambians' senses of increased accidents and threats to their general safety made clear that the majority of the population did not experience what they had been promised. In fact, the Copperbelt environment made hash of analyses based solely on standard economically-based indicators by incorporating potent elements of lived experience that Western policy manuals routinely left out. One of the most important elements had to do with supernatural experiences, some of which have already been evidenced in the ways that Zambians accounted for misfortunes. The strength of the supernatural to affect everyday lives became more obvious when Zambians considered the concept of good fortune in terms not only of on the job safety but of overall well-being and economic prosperity. In their experience, success could not be assumed to come as an outgrowth of hard work and of playing by a particular set of rules based on social contracts. Many times, prosperity appeared to be the result of harnessing dangerous forces that, in contemporary Western terms, had no exact counterpart.

At this point, however, at least a couple of cautious observations need to be made. First, the overall topic of supernatural experience, especially experiences of phenomena associated with witchcraft, is daunting terrain for Westerners to approach. Major studies conducted regarding African witchcraft in relation to Western thought patterns have been conducted for some time.[4] Yet, numerous sympathetic outsiders continue to note the inability of various academic disciplines fully to accommodate the subject.[5] Second, such an admittedly hazardous exploration cannot be made more manageable by excising and analyzing witchcraft as if in isolation from a larger context.[6] This is especially important for Americans to remember since the topic can seem appealingly foreign or exotic despite our own country's comparatively recent experiences of it.[7]

Copperbelt Experiences of Witchcraft

The intersection of inequitable or threatening situations with the phenomenon of witchcraft in Africa has been noted by other external observers.[8] Zambia's history also contains tense and complex relationships between those subject to witchcraft experiences and civil and ecclesiastical authorities. Witch-finding episodes, for instance, have occurred within various African Christian communities but, not limiting themselves to what some would consider a realm of private experience, have also had impacts on the country's public governance.[9]

On the Copperbelt of the early 2002 there also appeared to be two faces to this broad topic, one public and the other private. The public side emerged in the media and other generally accessible forums such as public meetings. Rather than dismissing witchcraft altogether, these discussions sometimes presented it as a pitiable remnant of outmoded thinking. At other times witchcraft experiences could simply be noted as part of "the facts" of a particular event.[10] Witchcraft was also occasionally portrayed as a bizarre, but limited, threat to the population as newspaper headlines such as "Suspected Witches Find Refuge in Kaleni Hills"[11] and "Traditional Healers Warn Witch-Finders Against Theft"[12] suggested. These public discussions of witchcraft could often be found juxtaposed against information pertaining to various concerns familiar to the Global North. Thus, the same *Times of Zambia* page containing an article on a flying witch also covered subjects ranging from global trade, to local manufacturing, to disability and housing concerns. The *Sunday Times* once ran a story entitled "Ritual Deaths Rock Chiawa" next to articles entitled "Environmental Awareness Centre Construction Completed" and "ZESCO Delays Banana Project."[13]

A public stance that acknowledged the existence of witchcraft while also seeming disengaged from it was quite understandable considering that Zambian criminal law declared witchcraft illegal.[14] From an official standpoint witchcraft might be a cultural factor, but it was not a political factor. This way of accounting for the phenomenon complicated outside attempts at seeking witchcraft's private face. Those whose livelihoods might be imperiled by hints of witchcraft—such as currently employed miners—were usually hesitant to talk of it. So, more candid discussions of witchcraft only seemed to emerge in situations where trusting relationships with outsiders could be built up over time.

Zambians used three terms: *muti*, *juju*, and *witchcraft* virtually interchangeably to signify the use of supernatural powers for manipulating or controlling a person's environment.[15] Frequently, the desire to manipulate or control someone or something seemed to arise from anxiety or malice.[16] A person might, for instance, spitefully use *muti* against another when motivated by jealousy.[17] On the Zambian Copperbelt of the mid-2000s, such jealousy particularly arose when one person or family appeared to enjoy financial success beyond that of everyone else.[18] This was so common that even the word "jealousy" could be used as a euphemism for "witchcraft" and to connote the means by which a person achieved material wealth and/or career success.[19] If someone said, "jealousy is there," for example, the phrase seemed intended to communicate that the person under discussion used witchcraft.[20] If a shopkeeper enjoyed great local patronage, the community might suspect that he or she used *muti* to boost sales.[21] Or, if an individual achieved success in his or her career, similar charges could be made.[22] During the course of this study rumors began circulating that a Zambian academic colleague of mine was a wizard. This accusation allegedly stemmed from his having grown up barefoot in a small village yet going on to receive an advanced degree from a Western university and becoming quite respected in certain learned circles. Despite his being an extremely capable scholar and administrator, he ultimately became the target of an informal campaign to encourage his departure from the institution where he taught. When he eventually did leave, his successor (who had an advanced degree from an American institution) had traditional cleansing ceremonies performed in the house before moving in.

Significant economic prosperity was also assumed to come to Zambians who associated with non-Zambians, particularly white people, and this could be hazardous to their Zambian relationships. *Muzungus* (whites) who were occasionally seen at township shops or traditional medicine stalls might influence the fates of their black associates in a couple of different ways. For one, the general public might conclude that what these proprietors offered was so good that even white people patronized them. Hence the public would patronize them, too. On the other hand, people could conclude that white patrons in the neighborhood were drawn there by the shop owners' *muti* use.[23] This could cause them to be ostracized.

An acknowledgement of witchcraft's double-edged force was matched by a widely understood sense that it had limitations. Witchcraft only *worked* in particular circumstances and when certain rules were followed.[24] Commonly accepted guidelines stipulated the barriers to *muti's* effectiveness and the circumstances under which it could properly be invoked. It was, for example, generally conceded that *muti* was ineffective on white people.[25] Several reasons seemed to account for this, including that successful *muti* use had to be done at close range between those known to each other.[26] Since the majority of Zambians never had occasion to associate with white people, the close range pre-requisite was seldom met. Even though, by the early 2000s, witchcraft was encroaching on urban areas where populations had a greater likelihood of mixing, Zambians knew that it was hazardous to try something on someone they did not know or even on someone from another tribe. The risk was that the intended recipient might have a more effective supernatural battery of his or her own.

> [I]n town it's only now that people are experimenting. But then, you'd never try on someone you don't know. Someone who is not your tribe, someone you haven't known for a long time. You'd just say, "No. You don't know this person so don't try because you may go there and find he's got more arsenal than you do and then you're in trouble." So if you are a new person in the compound or whatever, they would really just observe you until—so people are careful. So *muzungu* is too foreign for an African to even think of trying. You know, they just think, "Ah, it wouldn't work."[27]

These sorts of barriers could have acted in mine management's favor during the 2005 spate of accidents; for despite the general populace's anger with the mining companies, their managers were probably too physically and culturally removed for *muti* to have been an effective retaliatory strategy.

Muti also was not to be used on whim. There were certain rules and guidelines that had to be followed before invoking it.

> [T]he goodness of our witches or these people who bewitch people—whatever name we can call them—the goodness is that, especially where I come from, there has to be what we call *chitela*. . . . *Chitela* is "reason"; you can't bewitch somebody without a reason. Witchcraft, in its original sense, has got rules. It's like in war or in foot—or in boxing, you can't hit your friend below the belt. So even witchcraft, you don't just wake up today and just say, "Okay, I'm going to kill that person." No. There must be *chitela*. There must be a reason. So people in the villages lived—as long as you have never provoked anybody, you have never stolen from somebody's farm, you have never impregnated somebody's daughter and refused,[28] you have never done something ill to this family—you are almost safe. Now once there is a *chitela*, that magic works on you.[29]

Had the 2005 spate of accidents continued, the general public might have begun viewing them as accumulating into a *chitela* on behalf of the citizenry and against the mining companies. But, whether anyone eventually tested the rules exempting whites and those physically removed by trying witchcraft on company managers was never apparent.

One common reason for using *muti* in Zambian workplace settings was said to be the prevention of others' success and/or to ensure one's own. Worries about *muti* could affect how Zambians treated offices, dwellings, vehicles and equipment. Church bishops, for example, were known completely to change office furniture and houses occupied by their predecessors even when such moves came at great expense to already strapped diocesan budgets. Similarly, Zambian drivers feared to get behind the wheel of a vehicle that had been driven by someone they did not trust. And there was a common expression that workers in acting capacities or moving into jobs previously held by other Zambians would "sit lightly" in case the earlier occupant had been given to *muti* use. *Muti* could further account for non-productive workers remaining in jobs despite their evident incompetence. Such workers should be retrenched, people said. That they were not could only be explained by possession of some powerful *muti* used to protect themselves.

Muti or witchcraft was further credited as the reason certain individuals maintained unusual amounts of power in national life or the workplace. Such individuals seemed to have *muti* that protected them from questioning by others. In these instances, witchcraft apparently worked by making even the strongest, most forthright opponents weak at the moment of confrontation. How else could such determined people account for their becoming tongue-tied in the presence of those whom they had vowed to take on?[30]

The best *muti* was available to people in their home areas, so when politicians and church leaders spent time in their villages, popular talk was that they had gone to be "cooked." Kenneth Kaunda, for example, was apparently cooked to the point of frying in his home area of Chinsali. This was evident in his grayish skin tone, peculiar diet, and ubiquitous handkerchief, understood to be where he kept his *muti*. Frederick Chiluba's Congo connection assured him a good cooking.[31] And Levy Mwanawasa's visit to his mother's village shortly after assuming the presidency cooked him as well.

Upon occasion witchcraft beliefs or symbols could be used with effect across cultures. Not long after privatization, for example, an expatriate manager in one of the mining company corporate offices took advantage of witchcraft fears to guard his belongings. The incident began when he had miniature coffins made to accompany a Halloween costume. Subsequent to the party at which he had appeared in mortician's garb, the manager began using his baby-sized coffins to ensure against theft at his home and in the office. When going on holiday he would place one coffin at the front entrance to his house and one at the back and had no problems with break-ins. At one point, he also had thirty or forty boxes of personal goods that he wished to store temporarily at work. Because the boxes were visible through a window, he brought in the coffins and placed them prominently near the boxes. Zambian workers in the office complained and eventually took their concerns to the police. The expatriate manager then put the coffins in his car but knowledge that they remained on the premises still worried Zambians at the mining office. Their complaints continued until the manager was ordered by senior executives to remove the coffins completely.

Another well-known cross-cultural case concerned a European woman in Luanshya who took up with an African. Zambians of European descent recalled that the incident took place in the 1960s when the town clerk's wife left her husband and two small girls to live with their former gardener. The husband and children returned to the UK and he left her an airline ticket in case she should change her mind. She did not, however, but stayed and had children by the Zambian man, living like a woman of the townships. At one point, she had a little job serving at the cold meat counter of a local European shop, but lost this when the other white ladies complained. There was ultimately some question about the former gardener's faithfulness and general acknowledgement that the European woman's later years were not easy.

The Zambian version of this story included details that the gardener prepared a special tea to ensure his employer, the European husband, would like him. Somehow, the woman of the house drank the tea instead and fell in love with the gardener. By some accounts she then insisted upon marrying him despite his protestations over race and class barriers. Eventually, the couple did marry and the white woman adopted the simple lifestyle of a local resident. She could be seen bicycling about town and children from the marriage attended the local schools. By the mid-2000s, although the European woman and her gardener-turned-husband had died, a daughter was said to be still living in the area.

Malleable Material Boundaries

Manipulating or controlling one's environment by using the physical objects of *muti* or witchcraft was not, in Zambian experience, the only way that people could help ensure their economic success over against that of others. They could also employ the assistance of supernatural creatures or assume—with the aid of paranormal powers—a different creaturely form themselves.

African cultures have a well documented history of such experiences and, in Zambia, the crocodile was often associated with magical forces that could defy the material world's boundaries.[32] Newspaper stories that frequently described peoples' unfortunate encounters with crocodiles inhabiting streams throughout the rural areas told only the public portion of a narrative Zambians knew had a potent, private side as well. This aspect of the story did not always draw a straight line between a crocodile sighting and someone's trying to secure monetary gain. But the implication would be high that such was the case. As a smelter incident illustrated, in the mining communities threatening crocodiles might make appearances in connection with job security concerns.

Interviewer: Imibombele mwalebomba, kuSCAW, limbi kwalipusana nabalya abakumine, in terms of ama accidents, mwalekata amasanso?

Interviewer: Comparing with the mines, what about accidents in Scaw?[33]

Mr. T: Amasanso ya lesangwa nomba mainly ifyo twasangile yaleba ya ma magic.

Mr. T: Accidents were found. But mainly we attributed them to magic.

Interviewer: Ifyapala shani?

Interviewer: Like what?

Mr. T: Okay filya ifya kweba ati, crane yaimya basket, pantu ama basket, tuyeta ati ama bin basket, then baisula furnace pakwebati inshimbi shiye mu furnace. Ba shidumbwishamo pantu panshi palya kukakapo sisal. So sisal ilya nga yapya kwisuka, baikisha amacheni, inshimbi shaya mu furnace, so kwisa isalapo ci lid. So kupresa ama electrodes, fyamba ukocha palya. Nga fya sungulika palya fyabafye amenshi nga polegi, kulaba ukufumyapo slagi, filya ifiko fisa pamulu. Kwaliba umungwala, kulya kumungwala mwabikako icimuti icibishi. So palya pene kuti alefumya umuntu filya fine, amonafye mulyamwine mu furnace, limbi ni n'gwena yafumamo, mwamona. So nomba ena mukutilimuka kabili palya apaba ama furnace ni pamulu, mukutoloka palya, esemukutoloka nipa nshimbi panshi palya, umuntu alemana. Limbi umulilo wa fuma fye ku furnace wa mwikata ku li ovalo nakuli chi epulo.

Mr. T: Okay, like crane lifts a basket because the baskets, we call them bin basket. Then they open the furnace so that the metals will go into the furnace. At the bottom of the furnace, you tie sisal so when that sisal burns, it opens. Then they release the chains to allow the metals to go into the furnace and then you close the lid. After this, you press the electrodes and they start to melt until they melt and become like water or porridge. Then you remove slag, that dirt that comes on top. There is a bar to which you put a fresh log. So as one is removing, he just sees a crocodile coming out of the furnace, you see. So out of fear, and considering the level where the furnaces are, one jumps on metal to the bottom and as expected one becomes lame. At times, fire comes from the furnace and catches his overalls and apron.

Interviewer: Ninshi fyalecitikila ifyamusango yo?

Interviewer: Why were these things happening?

Mr. T: Okay like ifyaba life, pantu Scaw muno muKitwe tamwali, yali ku Livingstone, abaishile nayo banensu aba Lozi, yalifye kwati yabo. Tabalefwaya umutundumbi awe. Tabalefyaya umutundu umbi. So nga baingishafye umutundu umbi awe filya, ninshi awe fya shupa, mwandi.

Mr. T: Okay, the reason was that we did not have the original Scaw here in Kitwe. It was not here but in Livingstone. The ones who came with it are our friends the Lozi.[34] It was just like theirs, they did not want any other tribe. They did not want another tribe. So if another tribe is employed, then there was trouble.[35]

Within the mining townships, casual gossip also could link the appearance of crocodiles with people who seemed overly ambitious. Enroute to a gathering in Chingola, a miner in his fifties told a car full of friends about a comparatively well off co-worker who had a mysterious encounter with a crocodile.

Mr. M: [B]alebomba filyafine mumiine ukufika apobacitile retire. Palyapene epo baishile luba ilyo line. Baluba na bantu bashininkisha ati, umuntu nafwa pantu Ing'wena naimwikata.	Mr. M: [H]e was working in the mine until he retired. That is when he disappeared. When he got lost people believed that the person[36] has died, the crocodile has caught him.
Interviewer: Baile kukamana?	Interviewer: Did he go to the stream?
Mr. M: Baile no mwipwa wabo kukamana.	Mr. M: He went with his nephew to the stream.
Interviewer: Mukwikate sabi?	Interviewer: To catch fish?
Mr. M: Katwishi ifyo baile mukufwayako. So kulya kwine baeba umwipwa wabo ati, naisa nsambe ko. So ekufula ifya kufwala basha babika. Ekuya napamumana, palya pamumana In'gwena ekubekata. So balelwisha filya balepunda, umwipwa wabo ayako ukuti abadonse. Cilya ati abadonse, aisamona uku [unintelligible] cilya icakunuma [unintelligible] yabebisha napamenshi. So ebele bayama bafwa, in'gwena yabasenda. Yafika palya futi baisafuntuka. So efyo baile ifyo fine. So umwipwa wabo kusendafye ifya kufwala. So palipita, ni three weeks yapitile po ayi? Eee papitile three weeks, ninshi necililo baliteka. Nalamilanga pamwabo apobaleikala. Pamupundu apa. Kanshi kuLuanshya kulya bayabatola balifye ubwamba, bafwelefye kwati amasaka fye so.	Mr. M: We do not know what they went to do. So right there he told the nephew, "Let me take a bath." So he undressed and put the clothes aside; that is how he went into the river. Whilst in the water, a crocodile caught him and he started struggling, and shouting. His nephew rushed to the scene and wanted to pull him out. Unfortunately, he only saw the backside of the crocodile pull into deep waters. So he concluded that, "My uncle has died, a crocodile has taken him."After some time, the crocodile emerged again.[37] So that is how he disappeared, just like that. So the nephew took his uncle's clothes and went home. After three weeks had passed, yes, three weeks, and the funeral had been conducted, the man reappeared. I will show you where he used to live. Here at the mupundu.[38] He was found in Luanshya naked. He was just naked, and dressed in sacks.[39]

There were other ways that Zambians believed they could change their physical bodies in efforts to enhance their economic well-being. One common way was to, in essence, split into two people who could then work two jobs and collect two salaries.

Mudala uyu alebombela kusteel foundry ali ni Crane Driver. Alekomboka elyo alemoneka talekomboka. Nomba kwaishile isa mudala umo uwa kuMalawi, ewo baishile ingishe inchito. So aisalolesha aleingilafye mulya; "Imwe ngeci cilebomba mu?" "Ati iyo niba Banda." "Te Banda uyu." Ati "Awe niba Banda." "Banda crane kuno." Chaisa ci crane, apa. Aka-na sana uwa kuMalawi. "Kabiyeni ku nganda muye mumone Banda." Ukuya fika kunganda, ukuya sanga ba Banda bali kun'ganda. "Banda tuleya kunchito." Ukwisafika, ukusanga cilya icimuntu mu crane. Bonse naba mona no kumona ati Banda uyo muCrane. No musungu asumina nokusumina ati, kwaliba magic.

This old man was a crane driver at the steel foundry. He used to knock off but again he looked like he never knocked off.[40] The company finally employed a man from Malawi.[41] So when he entered the plant, he looked closely and asked, "You people, what is that thing working in there?" They answered, "It is Mr. Banda." He disputed, "This is not Banda." They countered, "No this is Banda." They called out, "Banda, bring the crane here." The crane came. The Malawian refused totally.[42] "Go home and you will see Banda." When they reached at Banda's home, they found him at home. They said to him, "Banda, let us go to work." When they came back, they found Banda still in the crane. Finally, even the white man believed that magic is there.

While potentially dangerous and sometimes described as wicked behavior, this practice seemed to be driven less by greed than by desperation. Despite official reports of Zambia's wealth increasing via rising GDP standards as a result of external interventions, many ordinary citizens experienced increased personal poverty due to job reductions and curtailments in social services. Engaging in supernatural activity in order to realize some economic gain was, then, perhaps judged unwise but understandable.

In a bit of linguistic irony, the term for people who split themselves this way had two very different connotations depending upon the usage context. In popular parlance, people who split themselves in order to receive double salaries were talked about as being "ghost workers."[43] These ghost workers were known generally by their ability to labor non-stop; their silence when spoken to; or perhaps their appearance just as flickering lights down a far tunnel. In terminology familiar to the West, these sorts of workers might be understood as "clones" since they allowed one person to be in two places simultaneously.[44] Western development discourse also uses the phrase "ghost worker" but within policy making circles this term tends to designate non-existent or deceased people whose names, left on payrolls and hospital patient lists, contribute to systemic or organizational corruption. Money for teachers, for example, might be siphoned off into fictional accounts while drugs designated for genuinely sick people might be claimed on behalf of a fictitious person and then sold on the private market.[45]

These phenomena, like those involving blood sacrifices, reveal the complexity inherent in relating theoretical economic principles to everyday experience of those living in locales targeted for development activities. They further attest to the difficulties that local people might have in trying to account for their circumstance using means that would appear acceptable in a modern setting. It might be further reasoned that Zambians sometimes employed language most typically associated with forms of Western Christianity. So, while some magical act might be understood as related to witchcraft or wizardry, it might also be declared Satanic or devilish, as the miner discussing his neighbor-turned-crocodile concluded.

Interviewer: Fyalechitika last year?	Interviewer: Did this happen last year?
Mr. M: Mu2001.	Mr. M: In 2001.
Interviewer: So abantu muChingola baliishiba?	Interviewer: So do people in Chingola know?
Mr. M: Bengi sana abaishiba.	Mr. M: Many people know.
Interviewer: Ninshi baleikalila kwati abapina lelo bali abalisambilila?	Interviewer: Why was he living like a poor person when he was learned?
Mr. M: Abamonekako nga fino ndemoneka efilyako, nomba imifwalile fye yabo ukufuma fye nakunuma, very poor. Nomba umuntu uwasambilila elyo alebomba mumaini, ne cifulo bakwete cikulu.	Mr. M: Had he been looking like me, like this, at least, but his dressing just from the past was very poor. Now a person is learnt and works in the mines with a respectable post.
Interviewer: Mukumona kwenu icitwalo muntu, ukusanguka ing'wena, ninshi banonkamo?	Interviewer: In your view what leads a person to turn into a crocodile? What do they benefit?
Mr. M: Ayomano yabu Satana. Imilimo yakwa Satana efyo yaba.	Mr. M: That is devilish [thinking]. The devil's work; that is how they are.[46]

Obviously, whether or not ideas that Zambians expressed in the vernacular could be thought of in one to one comparative terms with English words would be a matter of interpretation that only Zambians themselves could accurately assess. But the existence of such understanding in general—particularly with its apparently close relation to issues of economics—would seem crucial for development policy makers to consider.

Luanshya Revisited

Luanshya's historical experience with the supernatural has already been explored. It is also important to record how such knowledge affected and was affected by development efforts of the privatization era.

At the time and from a geological and economic development standpoint, the Luanshya mine was no longer worth the trouble of keeping open especially when compared to the vast potential of Konkola Deeps. But it had been one of the first to be sold.

This sale, considered a scam by all candid accounts, was only the first of several highly controversial developments. Mining managers instituted the celebration of Hindu religious holidays in the workplace and buried images of various gods—including that of the highly revered Hindu snake—on corporate office grounds and within the mine itself. Local Zambians found all this quite offensive. One woman who worked in the engineering department at the time told how her boss was instructed to go to the corporate boardroom and remove his shoes during Diwali. Further, he had to receive some sort of mark on his forehead that he subsequently washed off in the lavatory. Miners who were required to plant the idols underground retaliated by urinating on them as a means of "neutralizing" whatever power the gods might have possessed. Local people were objecting, she said, because they were Christians and these idols—these foreign things being put in the land—were not of Christian origin. The tumultuous situation in town, however, was not widely believed until ZNBC covered it on television.

Not only was the introduction of foreign gods offensive to Zambian religious instincts, popular wisdom considered the act a tip-off that the new owner's ultimate success with the mine would be short-lived. Hindu gods were obviously wrong for the Zambian environment and Hindu owners taking their gods underground desecrated the space. Consequently, people in the workforce and larger community knew that trouble would ensue.

In 2002, the mine ceased operating ostensibly for financial reasons. By some accounts, the stoppage was so sudden that copper under production was left half finished and Luanshya plunged into a desperate economic crisis. When the government of Zambia eventually settled on another buyer in 2004, close ties with the former regime prompted popular speculation, echoed by some in managerial ranks, that the new company was just Binani under another name.

Even though the mine eventually reopened, the general populace continued to encounter difficulties. By mid-2005, privatization had become rooted enough in Zambia to reveal the uneven wealth distribution patterns that inevitably accompany free market practices. The uniformity that, for good or ill, had characterized mining township life under nationalization was giving way as some prospered more obviously than others. Various people were now rumored to be "selling their years" or sacrificing family members in exchange for material prosperity. Notably, these actions were not publicly deemed to be witchcraft, but were designated as Satanic activity.[47] The essential ingredients of such a Faustian-like process involved blood and sacrifice—either the remainder of one's own life or that of close kin. Killers were said to be taking their victim's body parts to use in ceremonies designed to bring them wealth.[48] In one instance, a young man's clothes had been found at the pony club while his intact body was recovered from a stream some distance away.[49] Further, it was claimed that results of these killings could be seen in the town's second-class, or industrial, area where new shops were springing out of nowhere and individual businesses suddenly appeared to have astronomically high sales volume. One man was said to have bought fifty minibuses all at once.

The problem with this sort of charm, however, was its temporality. Wealth gained by such means would last only so long and then dissipate. In the meanwhile, ordinary citizens were becoming afraid to go out after 6:00 PM and red bereted policemen patrolled the streets.[50]

Questions of Reality

All of the instances recorded above, as with the previous chapter's discussion of accident causation, could be said to turn on questions of interpretation. To skeptics, an understanding that such stories represented peoples' perceptions would be enough to rob the information of credibility. Yet, the radically different nature of how Zambians accounted for both good and ill fortune when compared with a dominant development narrative's treatment of such themes deserves scrutiny by policy makers interested in being well informed before acting. Some of the most pressing reasons for this are the following.

When development is offered as sets of policies and procedures intended to bring greater flourishing to people's lives, there seems something egregiously out of harmony when entire swathes of policy recipients' experiences remain out of bounds for consideration.[51] What does it say about dominant narrative influences on policy making and planning when study after study (of which the present is simply one example) document that a rich, powerful set of ideas and experiences exists yet remains qualitatively at odds with depictions and expectations considered acceptable within official or professional settings?[52] When there is simply no place even to put the concepts of the people for whom the programs are ostensibly designed, it seems time to ask pointed questions about what is really going on and who the programs are really intended to benefit.

A second pressing point concerns the opportunity presented to Americans and others from the Global North for contemplating ways of interpreting reality that we have not generally considered in a very long time. How, for example, does the manner in which we typically account for disparity and ensure success affect the ways in which we make meaning at the most basic levels? That is to say, how has the intellectual approach shaping development narratives also molded what is generally taken for granted in the Global North as constituting success, failure, flourishing, and misfortune? Growing recognition that the earth does not consist of two independent hemispheres—as this case testifies—presents a chance for substantive revisions of taken for granted ideas within dominant story lines.

Third, ethically sensitive people confronted by the prospect that their taken for granted notions of faith, freedom, and civilization may—when propagated elsewhere—not bring universal good but rather can encourage destruction and death have a particular burden here. How do we come to discern our responsibilities for situations that promote devastation in the name of the exact opposite?

More painfully still, how do we then assume responsibility for making necessary changes?

These are pointed, brutal questions that go to the heart of how we understand ourselves and others. So it seems fitting that the next phase of this exploration addresses issues of identity and personal agency, or being and acting.

Notes

1. Sachs, in *The Development Dictionary*, discusses this connection throughout.

2. On North-South flows, see Veltmeyer, *Illusion or Opportunity*, 10. See also M. Shahid Alam, "Global Disparities Since 1800: Trends and Regional Patterns," *Journal of World Systems Research* XII, no. 2 (July 2006): 37-59.

3. See, for example, the work of The ONE campaign, http://www.one.org/ (accessed March 25, 2009) and of Episcopalians for Global Reconciliation, http://www.e4gr.org/ (accessed March 25, 2009).

4. Classic studies are those of Jean Comaroff and John Comaroff, *Modernity and its Malcontents: Ritual and Power in Postcolonial Africa* (Chicago: University of Chicago Press, 1993); Evans-Pritchard, *Witchcraft, Oracles and Magic Among the Azande*; Gluckman, *Custom and Conflict in Africa*; Hallen and Sodipo, *Knowledge, Belief and Witchcraft*; Moore and Sanders, *Magical Interpretations, Material Realities*. See also the more recent works of Adam Ashforth, *Witchcraft, Violence, and Democracy in South Africa* (Chicago: University of Chicago Press, 2005) and Heidi Holland, *African Magic: Traditional Ideas that Heal a Continent* (London: Viking, 2001). The present study is particularly indebted to Gluckman and Ashforth.

5. Geschière has mentioned anthropology's general inability to accommodate witchcraft and occult studies. *Modernity of Witchcraft*, 12. Ferguson encountered Christian/witchcraft dichotomies similar to those mentioned in this study and said, "there was a literalness and a materiality to discussions about witchcraft that my anthropological training did not prepare me for." *Expectations of Modernity*, 117-88. See also Hallen and Sodipo, *Knowledge, Belief and Witchcraft*, 87 and Adam Ashforth, *Madumo: A Man Bewitched* (Cape Town: David Philip Publishers, 2000).

6. Anthony O. Balcomb, "Sex, Sorcery, and Stigma—Probing Some No-Go Areas of the Denial Syndrome in the AIDS Debate," *Journal of Theology for Southern Africa* 1, no. 125 (July 2006): 104-14. Balcomb maintains that, "witchcraft should not be seen in isolation from its location in a broader worldview that includes many other things besides it. This worldview assumes interconnectedness in the universe, the existence of a spiritual world inhabited by forces that are both malevolent and benevolent and that impact on the material world, a sense of acute vulnerability to these forces accompanied by a belief that they can be accessed and manipulated, and an ontology that defines personhood in a communal and not individual way."

7. See a discussion of the Puritan worldview and Salem witch trials in David Hackett Fisher, *Albion's Seed: Four British Folkways in America* (New York: Oxford University Press, 1989). See also Hallen and Sodipo's discussion of Western witchcraft and Lecky's observations about Western culture's declining sense of the miraculous. William E. H. Lecky, *The History of the Rise and Influence of the Spirit of Rationalism in Europe*, vol. 1 (London: Watts and Company, 1910).

8. Ellis and Ter Haar, *Worlds of Power*; Adam Ashforth, "An Epidemic of Witchcraft? The Implications of AIDS for the Post-Apartheid State," *African Studies* 60, no. 1

(2002): 121-43. See, among others, Ferguson, *Expectations of Modernity*. Fields, *Revival and Rebellion*; and White, *Speaking with Vampires*.

9. The most famous example of this type might well be that of Alice Lenshina Mulenga, a young woman who, after a near-death experience in the mid-1950s, went on to preach of having had direct communication with Jesus and angelic beings. She eventually founded the hugely popular and prophetic Lumpa Church, characterized by the use of indigenous song tunes, witchcraft and magic cleansings. The Lumpa Church proved threatening to both the mission churches and Northern Rhodesia's political leaders for reasons similar to those of the Watch Tower movement. In the early days of the Kaunda administration, the Lumpa Church posed such an outright threat to earthly authority that it was banned. Taylor and Lehmann, *Christians of the Copperbelt*, 248. See also Norbert C. Brockman, "Lenshina Mulenga Mubisha, Alice," *Dictionary of African Christian Biography*, 1994, http://www.dacb.org/stories/zambia/lenshina1_alice.html (accessed December 31, 2009).

10. One article, for example, offered no obvious editorial judgments, but merely recorded that a woman in her 80s who had been flying a magical airplane had fallen, naked, to the ground in an Ndola township. She was beaten by a mob and later taken to the department of social welfare. "Suspected Witch Crash-lands in Ndola," *Times of Zambia*, March 7, 2006, 2. The fact of media coverage suggests that the phenomenon was actively experienced in Zambian life. And Ferguson, who encountered witchcraft discussions "constantly," regularly experienced Zambians' hesitation to admit their fear of it, expecting him to disapprove of the entire concept. *Expectations of Modernity*, 117.

11. "Suspected Witches Find Refuge in Kaleni Hills," *Times of Zambia*, July 7, 2005, 9.

12. "Traditional Healers Warn Witch-Finders Against Theft," *Daily Mail*, December 9, 2004, 3.

13. "Ritual Deaths Rock Chiawa," *Sunday Times of Zambia*, December 5, 2004, 1.

14. "Witchcraft Act," *Zambia Criminal Law*, vol. 7, ch. 90, 507-12, http://www.parliament.gov.zm/downloads/VOLUME%207.pdf (accessed March 25, 2009). There has been anti-witchcraft legislation in place since 1914. But Kaplan comments beliefs were "so deep rooted that few cases ever reach the courts." *Area Handbook for Zambia*, 188. See also Taylor and Lehmann, *Christians of the Copperbelt*, 251.

15. *Juju* seemed to be used a bit more commonly in the Lusaka area and *muti* on the Copperbelt. Powdermaker saw something similar in her 1950s Copperbelt study observing, "Good and bad magic and witchcraft exist among all tribal societies. They are based on a theory of personal causation and a belief in a supernatural power in the universe which men can control through their knowledge of magic and the following of taboos." *Copper Town*, 41.

16. See Powdermaker's observations about this in *Copper Town*, 41, and Ferguson's, *Expectations of Modernity*, 118.

17. Ashforth found jealousy to be a primary motivator for witchcraft in South Africa. *Witchcraft, Violence, and Democracy*, 34. Moore and Sanders found jealousy as an impetus for witchcraft documented in an African mining study of 1940. *Magical Interpretations*, 6. See also Kaplan, *Area Handbook for Zambia*, 189.

18. More than four decades ago, Max Gluckman documented similar situations on which he comments, "The man who is too successful is suspected of being a witch and himself is suspicious of the witchcraft of his envious fellows. Among the Bemba of Northern Rhodesia to find one beehive in the woods is luck, to find two is very good luck, to find three is witchcraft." *Custom and Conflict in Africa*, 96. See also Ferguson, *Expectations of Modernity*, 118. Balcomb, using Ashforth's phrase "social jealousy,"

asserts that an environment rife with perceived injustice—disparities between rich and poor far more obvious on the African continent than in America, for example—assumes that the envied one is the cause of the other's deprivation. Such an environment also makes the use of *muti* for both success and revenge more plausible. "Sex, Sorcery, and Stigma," 110-111. See also Ellis and Ter Haar, *Worlds of Power*, 114-40.

19. This would appear to recall elements of episodes recorded by Taylor and Lehmann. On several occasions in the 1920s and 1930s, witch-finding outbreaks associated with the Watch Tower movement involved well dressed young men from Nyasaland "which was regarded as the home of education and high wages." *Christians of the Copperbelt*, 26.

20. Ferguson also found these terms "used interchangeably." *Expectations of Modernity*, 119.

21. Powdermaker reported that, in return, successful people could be suspicious of other's spiteful witchcraft. *Copper Town*, 42.

22. Kaplan also acknowledges the means by which the use of *muti*, or medicine, could cut either way. "Sometimes the same medicine fulfills both [good and bad] purposes in that it protects its user by inflicting harm on someone or something else." *Area Handbook for Zambia*, 187.

23. Witchcraft fears resulting from association of Africans with whites have been well documented for quite some time. Geschière, *Modernity of Witchcraft*, 157. Similarly, my research assistants both spoke of receiving numerous negative innuendos and personal slights for their close association with a white person during the course of this study.

24. See Powdermaker's discussion of white and black magic. *Copper Town*, 41.

25. Hugo Hinfelaar also found *mu-sungu* being understood as a person not quite involved with the circle of life whereas *mu-ntu* was at the center of the circle. Hugo F. Hinfelaar, "Witch-Hunting in Zambia and International Illegal Trade," in *Imagining Evil: Witchcraft Beliefs and Accusations in Contemporary Africa*, ed. Gerrie Ter Haar (Trenton, NJ: Africa World Press, Inc., 2007), 229-45.

26. Michael Kirwen also found that, to be effective, witchcraft had to be used in close circumstances between people known to each other. Michael Kirwen, *The Missionary and Diviner: Contending Theologies of Christian and African Religions* (Maryknoll, NY: Orbis Books, 1987), 31. Kaplan observed that the effective range of witchcraft use was generally confined to family relations. *Area Handbook for Zambia*, 188.

27. Conversation with research assistant, 2005. Gluckman witnessed witchcraft being adapted to town life back in the 1960s. *Custom and Conflict in Africa*, 101. Ashforth, *Witchcraft, Violence, and Democracy*; Geschière, *Modernity of Witchcraft*; and Moore and Sanders, *Magical Interpretations*, all discuss various aspects of the phenomenon's adaptation to modern thinking and urban living.

28. I.e., refused then to marry the girl.

29. Conversation with research assistant, 2005.

30. Explanations that attribute unfortunate situations to the "how else?" of witchcraft have been documented and defended as legitimate reasoning processes by Evans-Pritchard, *Witchcraft, Oracles, and Magic*; Horton, *Patterns of Thought*; Gluckmann, *Custom and Conflict*; Mbiti, *African Religions and Philosophy*; Ashforth, *Witchcraft, Violence and Democracy*, and others. Balcomb, discussing witchcraft and HIV/AIDS in Southern Africa, notes that linking AIDS deaths to witchcraft can be "a matter of reason," since people dying in the prime of their lives otherwise simply does not make sense. "Sex, Sorcery and Stigma," 112.

31. Chiluba's background was Congolese and the Democratic Republic of Congo is, for Zambians, understood as particularly rich in witchcraft and wizardry.

32. Crocodile references are quite common on the Copperbelt. The reptiles inhabit many streams and local newspapers carry fairly frequent accounts of crocodile attacks on unfortunate citizens who happen to be in the water at the wrong times. See, for example, "Chilubi Couple Suspected as 'Human Crocodiles' Killed," *Zambia Daily Mail*, February 7, 2006, 2. In traditional Bemba understanding crocodiles were important symbols of ritual and royalty. Lane, *Moment of Encounter*, 31-45; Fields, *Revival and Rebellion*, 44.

33. Scaw was an Anglo American subsidiary company producing supplies for the mines.

34. The translator notes that the Bemba contains an implication that the Lozis were the one causing these deaths.

35. Conversation with research assistant, 2004.

36. I.e., the man who is the subject of this conversation.

37. The translator notes this refers to a widespread belief that, when crocodiles claim their prey, they submerge themselves then re-emerge to show what they have caught. Therefore, the speaker refers to this belief as a way of saying that the man was really taken. He was seen twice. Taylor and Lehmann observed quite a number of stories that involved people returning from death. Usually the purpose of such events was to bring a message to the living. *Christians of the Copperbelt*, 286.

38. Mupundu is a tree common to the area.

39. Conversation with research assistant, 2004.

40. I.e., he would finish his shifts yet it always appeared he was working.

41. The translator notes the reference to Malawi is an allusion to the widespread understanding that expert witches come from that region. Therefore, such an employee would be the only one who could figure out what was really going on. Banda, a real employee, had essentially split himself into two people in order to work two shifts and collect two salaries. See also references to Malawi/Nyasaland and the prevalence of witchcraft and witch-finding in Taylor and Lehmann, *Christians of the Copperbelt*, 26.

42. I.e., disagreed or denied.

43. There is some local historical precedent for such talk. Workers sharing jobs, changing names within the mine in order to work more or different shifts were commonly reported practices in the colonial era Copperbelt environment. White, *Speaking with Vampires*, 283.

44. This sort of supernatural manipulation of human bodies seems to bear some affinity to stories that White observed accompanying the colonial era mining environment. For example, stories circulating in Luanshya during the 1930s suggested that Europeans with special government licenses could employ Africans who worked in secret, assaulting coworkers with special medicine that would steal the worker's minds, allowing them to be taken anywhere. *Speaking with Vampires*, 270.

45. "$25 Billion Question," *The Economist* (July 2, 2005): 24-26. See also Ellis and Ter Haar, *Worlds of Power*, 123.

46. Conversation with research assistant, 2004.

47. Ashforth encountered similar circumstances in South Africa. *Witchcraft, Violence, and Democracy*, 41, 65. See also Ellis and Ter Haar's discussion of this phenomenon throughout sub-Saharan Africa and in the context of understanding money and wealth. *Worlds of Power*, 114-40.

48. Similar examples abounded in popular discussions over the course of this study. In mid-2004, a senior staff member at Mindolo Ecumenical Foundation preached a rambling chapel sermon in which he told a story currently circulating in town. It seemed that

a local man had died and been buried at Chingola Road Cemetery. A few months later, his family "forgot" that he had died and the man was spotted cleaning cars in Dubai. When asked how he got there, he told his inquirer that he did not know; he had just found himself there cleaning the cars. As the preacher's audience laughed somewhat nervously, he pronounced that sort of thing "Satan's work." (Dubai was frequently mentioned in connection with stories of sudden, surprising wealth. It was unclear whether this pertained to its exotic nature as a Muslim area or to its reputation as a place where many goods could be obtained.) In the early 2000s a church leader of my acquaintance lost his daughter, aged two or three, when she drowned in a bucket of water. Popular opinion stated that it was not really possible for a child of that size to drown in a bucket of water. This led to suspicions he had sacrificed her in order to become a more prominent ecclesiastical official.

49. There appears to be a strong resemblance between these sorts of stories and those that White reports from the colonial era mining environment. For example, vampires in the Katanga region were understood to be Africans who would capture other Africans in order that Europeans might eat them. Often the kidnappers left the victims' clothing behind. White further comments that the flourishing clothing trade between the Belgian Congo and Northern Rhodesia at that time would have made sightings of mounds of abandoned clothing all the more "chilling." *Speaking with Vampires*, 293.

50. See White for a similar instance recorded in late 1950s Kabwe (Broken Hill) in which evening youth activities were suspended out of fears that going out at night would invite kidnapping and murder. *Speaking with Vampires*, 219.

51. See Deneulin, *Religion in Development*, 5 on this point. That this situation is not coincidental may be surmised from the history of development policy itself. For example, the Basic Needs Approach first put forward as part of a United Nations initiative in the 1970s "explicitly recognized that development would not eliminate hunger and misery, and that, on the contrary, it would surely worsen the levels of 'absolute poverty' of a fifth, and probably of two-fifths, of the population. The Approach proposed the idea of dealing directly with the task of coping with those needs, instead of expecting their satisfaction as a result of the process of development." Esteva, "Development," 15.

52. While accounts of supernatural activity might have made it into public conversation via the Zambian media, Zambians almost invariably discounted and dismissed such stories in situations where non-Zambians, particularly whites were present.

♀

Chapter Seven

"We Are Not Slaves": The Pain and Power of Zambian Identity

For the funerals following the plant explosion, BGRIMM created a graveyard on its property within eyesight of the Kitwe-Chingola road. Red clay mounds typical of Zambian graves were interspersed with the charred remains of tree stumps and grasses that had been cleared to make way for the dead. Despite the public's doubts that accurate body recovery and identification could ever be achieved, a name marker adorned each grave. In front of this array was a small, oddly shaped brick structure that could have been intended as a monument. Placed at the foot of the brick heap was a hand-lettered sign beginning with the question, "Are we investing?" and concluding, "We don't want the Chinese. They don't care about our lives. We are not slaves."[1]

There is something remarkable about this image: an expression of international political and economic significance resting at the base of a monument that would probably never see completion nor be viewed by much of anyone save those directly affected by the tragedy.[2] It was the type of experiential commentary that development policy makers and corporate officials would have benefited from hearing. But the chasm between recipients of enacted privatization policies and those who designed them in theory was too great for much genuine encounter to take place. So, in a move reminiscent of leaving one's cares at the altar, the unknown mourner had simply submitted a great cry to the cosmos.

The sign's wording—a combination of suspicion, vulnerability, and resistance—indeed revealed much about the pain that surrounded the experience of privatization for a huge percentage of the population. This was pain felt as an overlay on (or perhaps interpenetrated with) the scar of colonialism, despite the country's forty years of political independence.

While the world of development policy making generally does not take into account the inherited pain of traditionally abused societies, there is a particular benefit to considering what those scrawled words at the BGRIMM site signified about basic identity issues. For identity, as should have been made clear in the context of Truman's Point Four Message, plays a crucial role in development policy making and practice.[3] Further, questions about identity in Southern African can only adequately be addressed in the context of colonial history. This is because of the manner in which Africans were assigned personas, characteristics, in essence, a distinctive group identity by outsiders.[4] Subsequently, descriptions of African societies, even names for various African peoples and their languages, were propagated internationally by those who had the technical means, such as printed words and mechanized modes of communication, for doing so. The legacy of these descriptors—particularly the extent to which colonial era depictions created impressions of Africans as being profoundly and negatively Other—can be detected in the age of development's early history. This is especially obvious in the Point Four program's assignation of categories of "developed" and "under developed" across the globe.[5]

Because of this history, effective contemporary counters to dominant development narratives need to involve searching, potentially uncomfortable inquiry into the ways in which all parties involved perceive themselves and others. Elements drawn from Zambia's history and contemporary mining community experience can be instructive here.

Roots of the Identity Struggle

Even in the days of Northern Rhodesia, Africans and Europeans who lived in Zambia talked about each other and themselves in ways that revealed both groups' reciprocal concerns. A dominant commentary implied that whites were stronger and more competent than Africans who were, by nature, weak in intellect and initiative.[6] The African Watch Tower movement, for example, attributed Europeans' wealth to innate wisdom granted them by God but hidden from Africans.[7] Anthropologists associated with the Rhodes-Livingstone Institute, a Southern Province-based research center in the 1940s, documented widespread assumptions that whites embodied appropriate standards by which to analyze and organize larger society. This view prevailed despite black Africans being the clear majority population.[8] Even as Britain was about to hand over Northern Rhodesia to the Zambians in the 1960s, the overwhelming majority of Europeans living there then thought Africans incapable of properly managing the nation and the mines.[9]

What the general population of Zambians thought at that time is a bit more difficult to discern from the outside given the scarcity of recorded first hand accounts. But Kenneth Kaunda's essays might well have summarized a general viewpoint. Shortly after independence he said that colonial efforts to destroy African self-confidence had disoriented the populace to such an extent that they experienced "a sort of walking civil war"[10] in their minds. Were Zambians Westerners? Were they Africans? Were they both or neither? The imprint of this sort of struggle was also evident in Kaunda's early policy making efforts. He declared, for example, that the new nation's government should be more "humane than efficient."[11]

But the mines—the government's chief source of income and the country's premier employer—were held to a different standard. These were to be run by disciplined, responsible, and vigorously active Zambians.[12] Kaunda's program of nationalization that eventually melded industry and government interests also evidenced identity issues, for the policy initiative was unabashedly enacted as a way of excluding foreign interests in the country. Taken in context, these policy positions and actions could appear reasonable—or at least comprehensible. Yet identity dilemmas eventually inundated ZCCM. Was the mining sector a capitalist industry running on military lines as experts and professionals insisted must be the case? Was the mine company the government's main social welfare program as local people who depended upon it had come to expect? Was ZCCM the premiere means of defining who Zambians were as a nation? Was it merely a collection of civil servants tending to a cash cow?

These questions continued to be debated in the re-privatization era when a new generation of expatriates contributed additional perspectives to the situation. It would be inaccurate, however, to suggest that this group of outsiders maintained exactly the same attitudes as their predecessors. For one, the context within which expatriates and Zambians began working together in the mid-2000s was clearly different than before. By the time re-privatization was fully realized, Zambia's several decades of political independence had helped secure increased opportunities for Africans to experience cross-cultural interactions.[13] New generations of Zambian miners had also grown up amid international civil rights movements. The overall mobility fostered by globalization further meant that the country's re-privatized workplace was substantively different from the colonial and early independence eras.

Still, what could be considered a proper relationship between the mining companies and Zambians was a contested question, evident in contrasts between mining regions and the rest of the country. While the mine sites underwent rebuilding on a scale unmatched since the colonial era,[14] other aspects of existence such as peripheral roadways, houses, and recreational facilities were no longer considered company concerns. Jettisoned by the corporations and neglected by a government known for its shallow depth of expertise at managerial levels, the country's industrial and social infrastructure remained generally bedraggled.

Despite all this, from the expatriate perspective Zambians could be faulted for inaction. They did have money enough in their pockets to buy beer. So what inhibited them from initiating even a few small improvements on their own? It was clear from the ways in which they talked that no one liked how things looked or functioned but it was also clear that the days of Mother ZCCM's oversight had passed. Moreover, since the re-privatized world intentionally rewarded innovation, what was preventing more Zambians from acting?

There was also evidence of a highly contested relational context among socio-economic groups of Zambians themselves. One vignette from the field work of this study can illustrate. As my research assistants, several miners, and I sat one afternoon at a tidy outdoor bar near the Mufulira mine site, we encountered Oliver, a former roommate to Enock, who had since become the owner of a small timber business. Oliver told our small group how he had begun his business with only ZMK40,000 (about $10 USD at the time) and now had 18 employees whom he paid a competitive wage. Soon the talk turned to entrepreneurship. Various people wondered what inhibited more Zambians from following

Oliver's example and what the country could do to build infrastructure for a better business environment. This led to thinking about Zambia's tax collection method, a cumbersome and inconsistent affair that required taxpayers to appear at one of only a few sites scattered throughout the country. Many affluent Zambians escaped taxation simply because it was either too much trouble or impossible for the government to locate them. As the Mufulira group delineated the enormous underlying complexity needed to keep everything functioning, Enock became increasingly agitated. Finally, he stood up, declared, "We are bewitched and the one who has bewitched us has died," and walked away. The Zambian saying to which he referred connotes a traditional understanding about witchcraft. If one is bewitched, it is necessary to go get the person who worked the spell in order to reverse it. If that person has died, however, the spell is impossible to overturn.

A theoretical discussion about this matter would likely turn first to managerial and operational steps that would be necessary for Zambian policy making and practice to move from its haphazard tax program to one more broadly representative of and beneficial for society. Yet why did Enock react as he did when the topic arose? When considering that Zambia was entitled to more revenue than it took in as a modern nation state, he could have offered any number of economically, historically, or politically related commentaries. He could have contended, for example, that inefficient government processes are ultimately no match for free market innovations, so an environment that bolsters the common good should be created by private rather than public means. Or he could have expounded on African urbanization trends and the comparative advantage of seeking taxes from rural and urban dwellers, small and large-scale businesses. But he chose instead a vivid commentary that spoke to something else entirely. When it came to creating a vibrant economic environment, Zambians collectively were stuck just as surely as they would be if individually searching the bush for a deceased perpetrator of magic spells.

Regardless of whether Enock used witchcraft imagery literally or figuratively, his observation was quite relevant to some very basic matters. "We are bewitched" implies a life condition determined by something other than the ones affected while "the one who has bewitched us has died" suggests an inability to do anything about the situation. Ultimately, questions such as "What determines who I am?" and "What am I capable of doing to affect my situation?" are questions about identity.

Under the guise of development policy making it has been comparatively easy for outsiders to pose identity-related questions about Africans for quite some time, although these may be couched in terms of good governance, anti-corruption, or self-sufficiency.[15] In the Copperbelt case, tensions concerning identity and ability to act arose constantly, often complicating workplace life and highlighting disparities in understanding and meaning making. Enock's troubled remark hinted at that. So did many other comments and actions.

Dilemmas of Present-day Zambian Elites

Quite a few Zambians had trained under the mines' flourishing personnel systems during the 1960s and 1970s and a notable number had pursued education overseas or at the University of Zambia.[16] This stratum of mine personnel achieved a marked level of upward mobility through the mines' managerial ranks, creating a small but quite elite population that had not existed prior to independence. Although the privatization program attracted an influx of expatriates, their numbers were quite small in comparison to colonial days and they worked within an environment that included Zambian former ZCCM executives and managers. There were also numerous Zambian owned businesses that supported the mines by selling supplies and skilled labor on contractual bases. Along with their Western training and dress, members of the elite Zambian business group also frequently sounded sympathetically Western in their expressions and sentiments. Even when their fellow citizens' reputations were under scrutiny by outsiders, Zambian mining executives and managers often publicly voiced viewpoints that closely aligned with their expatriate counterparts. Many elites, for example, corroborated a public or dominant narrative that the nationalized mine scheme had failed because Zambians, stretching their expertise too thinly, had not reinvested properly in the mines. Further (as expatriates often maintained), it *was* inappropriate of ordinary Zambians to expect their more affluent relatives to provide for them and Zambian officials *did* too readily allow outside interests to determine the country's policies. Whether or not Zambian elites wholly believed these ideas in their hearts was another matter, however, for there appeared at the same time to be universal acclaim for the ZCCM era. This was especially true regarding the social benefits such as recreational clubs and township amenities that characterized the time and that, according to the rules of privatization, had helped distract ZCCM from its true mission.[17]

Additionally, whether or not Zambian elites were role models for subsequent generations, whether they embodied strong political and economic ideals of benefit to the country were contested matters. In some ways, this debate about motivations and models harkened back several decades to wage and labor negotiations (specifically the matter of equal pay for equal work) during the period of African Advancement. Researchers at that time found widely differing opinions among Zambians who worked strictly as laborers and those who had managed to compete on equal footing with whites for higher level appointments. In fact, it appeared that some officials from the African miner's union put forward the opinion that more elite Africans were only interested in themselves while the majority of laborers assumed that African Advancement would mean "much higher wages for all Africans."[18]

To younger Zambian observers of the mid-2000s, the stratum of elite managers who worked in positions of authority at the re-privatized mines had lost something of their Zambian-ness. They had, in fact, become essentially expatriates. Described as "Afro-Westerners"[19] by some and simply viewed as Western by others, Zambian elites were frequently suspected of not truly comprehending what it was like to come of age amid privatization and globalization.

Despite their telling stories of enduring and overcoming colonialism, Zambian elites had grown up in an environment of relative plenty. Plenty of jobs had been available even if the types of work were limited. Plenty of life's staples—if not the frills—had easily been obtained. These older Zambians, then, simply had not experienced the want and deprivation that typified life for most people in the early twenty-first century. So how could they genuinely understand the reasons why few ordinary citizens exhibited initiative or drive?

Township Initiatives

In the event that elites or outsiders criticized Zambians generally for lack of drive or personal agency the lives of township dwellers provided powerful alternate testimonies. Many Zambians did lounge on old, shredded furniture in dismally appointed bars. They also walked along garbage-lined roads on which the surface tar had long since washed away. But images such as these—the sorts that Western media often displayed for their audiences—did not capture the entirety of existence there. The mine townships had been constructed when Zambia's population was significantly smaller and when miners customarily departed town for the rural areas upon retirement.[20] In times past, the compounds, as they were sometimes called, had also been maintained first by the private colonial era companies and later by the behemoth ZCCM.[21] By the re-privatization era, though, far more people lived in the townships than their infrastructure had ever been designed to accommodate while maintenance and oversight were virtually non-existent. Additionally, township dwellers had to contend with the HIV/AIDS pandemic, increased malaria incidence, and poverty of an intensity unknown during the pre-colonial and colonial eras. Yet Zambians managed to conduct their lives with marked exuberance and resourcefulness. As evidence of this, interspersed among the old taverns and ranging along the dirty streets were numerous community schools flourishing in private back yards or abandoned buildings. Overseen by teachers who somehow managed to acquire schoolbooks, chalk, even the occasional light bulb for nighttime classes, these schools attracted scores of teenagers for extra lessons after hours. Small vending stalls or tuck shops—forbidden when the nationalized companies had controlled services—abounded on the townships' pot-holed roads, allowing informal sales of vegetables, soap, and cell phone cards to supplement family incomes.

Superficial exposure to examples of such resilience, along with the cheerful manner in which Zambians generally conducted themselves publicly, could be interpreted by external observers as evidence that Zambians either were unaware of the discrepancy that existed between themselves and residents of the Global North or that they were happy despite the disparity. In the mid-2000s, however, although ordinary Zambians could be stunningly gracious to outsiders, the majority seemed quite cognizant of their place in the global economy. They often displayed distrust for policy makers of all sorts, having seen so many convoys of Land Cruisers carrying visiting consultants come and go over the years. Further, Zambians generally did not appear to distinguish between colonial predecessors and the foreign investment corporations that had bought the mines in the late 1990s and early 2000s. Times of crisis, such as the 2005 spate of accidents, especially revealed a melding of past and present perceptions, allowing some public officials to insinuate to receptive audiences that "the investors" in the privatized mines were the same people Zambians had chased away at independence.[22] The citizenry, however, also partially faulted government leaders for letting colonialist intruders back into the country. Many seemed to say that the majority of Zambia's struggles could be attributed to *aba abaisa*—"those who have come."

So, overall, the contrasts between mining life in the ZCCM days and life under privatization, according to everyday Zambians, could hardly have been starker. In the old days, for instance, the mines had come looking for good people rather than good people having to wait for whatever scarce jobs might open up. In those days, Zambians had filled upper managerial ranks and compound life was pleasant. Retrenchments and massive poverty seemed only to have come with the return of white people.[23] Therefore, privatization was not really the *result* of earlier problems. It *was* the problem.

Language and Experience

Such private assessments, or alternate narratives, offered by ordinary Zambians represented a radical departure from dominant development narratives and could be accounted for in several different ways. Part of the problem lay with terminology. Mining understood in terms of "wasting assets" meant nothing to men whose fathers and grandfathers had spent their careers underground. How could it possibly happen that Zambia's famed copper mines could be depleted? That was simply incomprehensible. Part of the problem was also attributable to the ways in which information about the privatization program had filtered out to the general populace. For whatever reasons, the chronicling of which could constitute a separate case study, descriptions of life under a free market economic system received insufficient translation into vernacular terms and concepts. Consequently, many ordinary Zambians appeared to have believed that life in the re-privatized era would continue as before but with the fortuitous addition of their receiving more money. That privatization actually involved profit-driven retrenchments and elimination of services came as more than an unwelcome shock.

> You know, you try and tell this person that mining is a wasting asset. They do not put it in their heads. They think—they will tell you, "My father worked at the mines. And I have worked at the mines. My son is now working at the mines. What are you telling us that our mine will come to an end?" They couldn't just comprehend that. . . . Some of my family and my friends have lost jobs. And the impression that—you see, the people were not properly told what would happen when privatization comes. They thought they would continue with what has been happening before—what they were doing before. But what would change [was] that they would be getting more money. That's the impression that was created. . . . So that when people—at the time they were losing

> their jobs, they didn't realize that you see, in a private industry you want people who are hard workers. Committed. And skilled. And you want a leaner labor or workforce to work for you. So that you can make more profit because you don't want to lose part of your profit—waste it on people who are not contributing anything to the company. You see what I mean? So when people started losing their jobs they said, "Oh, is that all what privatization is all about?" You see then they started to [regret] it. Basically they think now that it's a demon—privatization. If you ask most of the people in the community they would say privatization is an evil thing.[24]

This perception was exacerbated by lack of access to economic and political power holders. So the Zambian citizenry had only a generalized portrait of those who might have been involved in creating the rules under which they now had to exist. The picture they consequently constructed was of policy makers—whites, no doubt—who were invulnerably detached from and uncaring about the situation of blacks.[25] Miners felt distanced particularly from the companies' expatriate bosses and from vague, faceless "investors" whose decisions handed down from locations unknown palpably affected people's lives. The detachment that Zambians experienced in this regard was evident in vernacular speech when, for example, they used terms for "the investors" and "visitors" interchangeably. As with its English counterpart, the local language equivalent of "visitor" denoted a person just passing through. Everyone knew that visitors were expected not to upset things during their temporary stay in a place, but these investor-visitors had. *They* were behind the sorrows of re-privatization and still they displayed little interest in making real commitments to the country. In fact, to underground workers in the privatized companies, there was even considerable mystery as to the origins of corporate policies. Whoever gave the companies' executives their orders must have been very far away. Workers suspected this because the company managers seemed to write and send reports yet nothing ever changed.

In sum, Zambians appeared to find expatriate managers and their investors uncaring and aggressive. They were not wanted or needed yet they dared to return to Zambia's mines where they cavalierly demolished the earth and the local people at the same time.[26] Why this was the case, according to popular Zambian interpretation, could only be attributed to selfish greed that also bred some fearfulness in outsiders.

> It's the fear. They knew that, "if the Africans became learned as we are, they will chase us." You see the tricky part of that? You see, this is what the white man has been scared of all the time. And that's why he's always afraid because he knows that . . . once the black man becomes stronger, he is going to destroy the white man! So as a result there is this, you know, you—maybe you can call it a Cold War—there is this inner thing which the white man has got—that fear.[27]

Having assessed expatriates' attitudes thusly, Zambians found external criticism of their own initiative and abilities to be particularly infuriating. Zambian miners countered any disparagement of their capacities to be effective employees with accusations of their own against expatriates. Managers in the newly privatized companies, for instance, rarely bothered to motivate workers. How could people who sat in offices all day or people who simply rebuked their workers expect the enthusiastic and energetic involvement of the labor force? This had not been the case in ZCCM's days. Moreover, expatriate managers who accused Zambians of reticence or poor initiative merely demonstrated how little they really knew about Zambian society, how little they saw of the overall picture. Expatriate managers especially had no idea of the double-edged dilemma that initiative itself posed within Zambian society. In hard economic times it could be not just difficult, but dangerous for Zambians to display too much internal drive. Starting small businesses or independent schools could be rewarding to the community but also quite risky to the individual if the one taking initiative were eventually judged to be outstripping the overall community's welfare. Whites might understand the concept of envy. But did they truly understand the ramifications of jealousy in the Zambian context?

Jealousy has already been introduced here. It has figured into perceptions of accident causation and material accumulation through its placement within a particular cosmological context that accommodates an active spirit world. Zambians seemed to acknowledge that phenomena such as witchcraft also helped define their views of themselves and their abilities to act within the material world. Jealousy, for example, could make the motivated worker's life miserable even if all that happened involved the petty sniping of co-workers. Such everyday sabotage could not strictly be classified as witchcraft but could have similarly devastating material effects. Jealous gossip was so common that Zambians could talk of it idiomatically. *Twala filial musenga* (literally "we will eat it in the sand") conveyed how the figurative desire to knock an envied one's accomplishments to the ground would at least allow others to partake of the broken and sullied remnants. Further, because Zambians knew that supernatural forces present in the workplace could be invoked for jealous purposes, the idea that they naturally possessed good senses of personal responsibility and initiative seemed even more reasonable. That is, attributing workplace accidents to short cuts or slack supervision did not take into account that the very possibility of workers being snared by *ifiwa* or succumbing to a *chisomo* attack would prompt more, not less, care on the job. Aberrant behavior of the type that managers might label "irresponsible" had another force or form of agency such as *chisomo*, *iminyama*, or *ifiwa* behind it.

Taken all together then, work under privatized conditions differed regrettably from what miners collectively recalled as an earlier era of great strength and pride. In the not too distant past, both labor and management had been dominated by Zambians. During that time, workers had earned reputations for their bravery and competence in the face of danger. And, as both public records and private recollections made clear, Copperbelt miners had consistently demonstrated fortitude for living resiliently. Their quick mastery of union style organizing back in the 1930s, for example, had helped spawn a self-assertive workforce unparalleled in surrounding regions.[28] Every Zambian knew that, from the colonial days onward, the collective conscience of the country and real strength of the populace resided not with Lusaka's politicians and privileged class, but on the Copperbelt among ordinary people.

Nurturing and maintaining a sense of distinct identity, then, involved collective, many faceted efforts. For individual families, both husband and wife understood that their separate roles played a part in keeping miners safe underground. For the workforce as a whole, keen knowledge of the rules and a willingness to engage in creative, collaborative actions helped keep laborers from being completely overwhelmed by corporate power. Public protests, for instance, might be led by miners' wives who, borrowing their husbands' boots and hardhats, would march outside head offices and plant gates. The mining companies could fire paid workers for disrupting production but what recourse did they have against their spouses?[29]

Many in the Zambian workforce had had their identities as miners reinforced through school books and popular songs that conveyed a sense of competence and accomplishment. Until some time in the 1990s, the same books used to teach Zambian students how to read had also relayed the adventures of rural men who found success working in the mines. These books were ultimately deleted from the government's education curriculum during a series of revision efforts funded in part by the World Bank.[30] Up until about the same time these educational changes were being made, popular songs and poems also went around exhorting national pride in the mines and the men who worked them.[31] Mine captain Teddy Chilambe, for example, sang a well known song that made clear the integral link between mining, identity, and the common welfare. It defended miners as the greatest workers in the country, directly crediting them for the development of Zambia's schools and hospitals.

In general then, while Zambians appeared quite aware of the many ways in which outside forces attempted to render a negative or disparaging portrait of them as people and as actors, Zambians understood that their lives were far more complex than any external definitions could accommodate. For one, a cosmos with a mysterious and powerful range of forces and actors rendered Zambians less autonomous or free to act than outsiders might have realized. For another, Zambians were aware of a past in which miners—by demonstrations of strong identity and agency—had won better working and living conditions for themselves. Access to a range of spiritual, historical, and experiential knowledge useful for countering present day forces was, however, increasingly hard to achieve. By the mid-2000s, in contrast to long standing Copperbelt tradition, many miners no longer encouraged their sons to follow their predecessors into underground careers.[32] Resentful and worried beneath polite and calm exteriors, many Zambians seemed to feel stuck in lives that rendered them unable to do much beyond struggle for survival.

Notes

1. In Copperbelt Bemba it is possible to have entire conversations in the first person while meaning the second or third person. "Are we going?" said to a friend in the company of an unwanted third party could well mean, "Is that other person going?" So, the question about investing or creating slavery, while phrased as if to fellow Zambians, may have been aimed at the foreign owners.

2. In fact, within a few years the gravesite was overgrown with weeds and barely visible from the road.

3. Haynes maintains that the Truman program was formulated over against the supposed backwardness of so-called undeveloped countries. Consequently, they needed "salvation" obtained through the application of capital and modern technology. *Religion and Development*, 2. In other words, undeveloped countries were identified as inherently unsuccessful—what contemporary slang might refer to as being "losers."

4. See, for example, various musings contained in Chauncey Hugh Stigand, *Hunting the Elephant in Africa and Other Wanderings* (1913, African Hunting Reprint Series vol. 8, Bulawayo: Books of Zimbabwe, 1989), 205.

5. Albert Memmi, *The Colonizer and The Colonized*, trans. Howard Greenfield (1st American ed. New York: Orion Press, 1965; Boston: Beacon Press, 1991). Basil Davidson, *The Black Man's Burden—Africa and the Curse of the Nation State* (New York: Times Books, 1992). Rist, *The History of Development*.

6. See, for example, Stigand, *Hunting the Elephant*, 157.

7. See the discussion of Watch Tower in chapter three. Watch Tower also contended that if Africans prayed fervently enough, God would send the Europeans home and leave their wealth behind. Lane, *Moment of Encounter*, 14. Note the prayer was not for God to make the Africans as wise as they perceived Europeans to be, but to allow assumption of the Europeans' wealth. Lane goes on to assert that, from the time of earliest European exploration, as "they appeared within the African world, Europeans were seen as beings from a transcendent realm possessing sacred power. On the other hand, as Europeans encountered Africans, they condemned them as being barbaric." *Moment of Encounter*, 37.

8. See the discussion of this situation in Ferguson, *Expectations of Modernity*, 105-06.

9. Wood, *Northern Rhodesia*, 62; National Institute for Personnel Research in Association with the Institute for Social Research, *The Attitudes of White Mining Employees Toward Life and Work on the Copperbelt* (Pietermaritzburg: University of Natal, 1960).

10. Kaunda, *A Humanist in Africa*, 31.

11. Kaunda, *A Humanist in Africa*, 31.

12. *Towards Complete Independence*, 5.

13. Beginning in the 1950s, for example, organizations such as the World Council of Churches tried to address inter-racial issues by holding camps where blacks and whites worked, ate, worshipped, and roomed together. The first such WCC effort in Central Africa took place in July and August, 1959 at Mindolo Ecumenical Foundation. "Mindolo Work Camp," (minute book, World Council of Churches, Kitwe, July 10-August 9, 1959).

14. According to a survey published by the Chamber of Mines, from the time re-privatization began through the latter portion of 2005, the mining companies had invested about $1 billion USD in plant rehabilitation, expansion, and new facilities. "Survey of the Zambian Mining Industry 1995-2004" (Kalulushi: Chamber of Mines of Zambia, 2005), 2.

15. Memmi, for instance, comments on Western bemusement at the African's apparent "lack of desires, his ineptitude for comfort, science, progress, his astonishing familiarity with poverty." *The Colonizer and the Colonized*, 82.

16. While five teacher-training colleges were established in Northern Rhodesia between 1953 and 1963, there were no universities until The University of Zambia (UNZA) was founded in 1966. P. C. Manchishi, "The Growth of Teacher Education in Zambia Since Independence," *The African Symposium* 4, no. 4 (December, 2004): 3-4.

17. The extent to which African elites working in the Copperbelt mining industry have shared the philosophical and ethical outlooks of European expatriates has always been a convoluted question. In his introduction to the second edition of Davis, Robert Rotberg notes that Davis observed in the early 1930s how the ability of some Africans to become fairly quickly sophisticated in European culture produced a double-edged result. On the one hand, it allowed Africans to become somewhat assimilated into the European world but is also allowed them to use their knowledge to push back against oppression. Rotberg in Davis, *Modern Industry*, xx.

18. Taylor and Lehmann cite this situation as general evidence of "inherent mistrust of the responsible leader who stands at the point of contact with the 'white' world, enhanced by tribal rivalries." *Christians of the Copperbelt*, 147.

19. Taylor and Lehmann talk about this phenomenon in terms of "Africans who have adopted Western categories of thought." *Christians of the Copperbelt*, 275.

20. It should also be recalled that African miners during the early colonial era were not allowed to live anywhere but in company owned housing if they resided in a mine African township. By the 1950s, some African freehold townships did exist on the Copperbelt outside of Kitwe. These were run by a District Commissioner in concert with an African management board and miners could build and own homes there. Taylor and Lehmann, *Christians of the Copperbelt*, 30.

21. It should not be assumed, however, that compound maintenance when performed by the private companies in the colonial days was faultless. For example, Davis observes that the influx of women and children as families moved to town to accompany the miners created social and hygienic concerns. *Modern Industry and the African*, 378.

22. At the height of the 2005 crisis, then Minister of Mines Kaunda Lembalemba was harshly critical of various mining companies while at the same time observing that Mines Safety Department officials were hampered from making their rounds due to lack of transport. Interview on ZNBC-TV, May 20, 2005. It eventually emerged that Lembalemba—who could be seen driving around the Copperbelt in a Mercedez-Benz—was being paid to denounce certain mining companies in favor of others.

23. Even though poverty and food shortages had begun to spread in the late 1980s, miners and their families were largely insulated at that time. The mines' abilities to buy products in large quantities and distribute them through their company stores meant that it was some time before the extent of the country's economic situation had an impact on its largest labor force. It was only by the mid- to late-1990s that problems began to be obvious in the compounds.

24. Geologist, Copperbelt mining company, interview with author, 2004. The Chamber of Mines report on the progress of privatization acknowledged this popular sentiment of privatization as well. "To most Zambians, the evil associated with privatisation of parastatal companies, the mining conglomerate inclusive has been the apparent or in some cases real loss of jobs in the formal sector due to operational rationalization exercises that the various privatised companies have undertaken." "Survey of the Zambian Mining Industry," 15.

25. See discussions of relational and informational barriers as contained in Fields, who also argues that the colonial project was essentially medieval, not modern, in structure. Nevertheless, elements of that infrastructure including missions and state bureaucracies largely dissipated at political independence. Consequently, it was impossible for Africans to construct anything similar without investments of huge amounts of funds. "The postcolonial regime that wills its society modern must pay the price of overthrowing and then replacing the institutions that made its predecessor possible." *Revival and Rebellion*, 274-76. This might, in part, explain the development dilemmas faced by President Kaunda as well as the need for ZCCM to engage in job fragmentation.

26. Interestingly, however, there were times when European or North American expatriates appeared to be preferred to Indian or Chinese investors as the incidents at Luanshya and Chambishi illustrated. Additionally, as one miner commented, "That is why that time when Vedanta has taken over, the accident *bruuh* everything down BGRIMM, eeh even that, that is Musonda falls accident and KCM you see, so because the way they came with those spirits, they had their own spirits, and concerned to the spirits which are within the mines there they can't tolerate things and with Zambians, especially we the people are annoyed that the mines have been given to the Indians. . . . How can they sell the mines to the Indians? We wanted . . . Americans." Interview with author and research assistants, 2005.

27. Former ZCCM draftsman, interview with author and research assistants, 2005.

28. Taylor and Lehmann record a series of developments and associations from at least the 1930s that they saw as "more indigenous and uninhibited organs of African self-expression." *Christians of the Copperbelt*, 136.

29. Parpart's study of women on the Northern Rhodesian Copperbelt documents the tensions of colonial powers wanting women "in the mine compounds to be loyal working-class wives, to increase labor productivity while making minimal demands themselves." Yet the authorities were also frustrated by women who were able to turn this situation to their own advantage when protesting corporate policies. Women were known to "support their men in the class struggle" by joining strikes, cooking for strikers and cheering them on. "Class and Gender on the Copperbelt," 10-11.

30. Manchishi, "The Growth of Teacher Education," 305. This may be an example of what can happen when outside policy makers think they are doing the right thing but fail to realize the interconnection between their actions and other effects. The World Bank began its involvement with the Zambian education system with four loans before 1980, canceling a fifth in 1987 in response to President Kaunda's withdrawal of involvement with the Bretton Woods Institutions. The Bank also undertook a project in 1992 to support "rehabilitation and expansion of primary schools and the provision of learning materials." "Zambia Country Assistance Review," 121. See also Hoyt Alverson, *Mind in the Heart of Darkness: Value and Self-Identity Among the Tswana of Southern Africa* (New Haven, CT: Yale University Press, 1978), 90.

31. See, for example, "Tribute to a Hard Working Miner," *Mining Mirror*, April, 1996, 5.

32. In the early 1960s, Jamieson observed that priority had been given to sons and grandsons of those who had first gone underground in the 1920s. This labor pool had been so deep that, combined with the lack of other industrial opportunities for Zambian laborers, the mines at that time did not have to engage in recruitment activities. *Zambia Contrasts*, 69.

♀

Chapter Eight

"They Are Always Suspecting Us": Expatriate Experiences of the Copperbelt

Although case studies can offer readers the benefit of vivid impressions of people and places that might not be gained otherwise, case examinations can also carry risks. Most importantly, in relating a narrative of a particular circumstance that involves disparate groups and interests, it is all too possible to caricature inadvertently some of the actors involved. Overlooking issues that would be obvious to those under consideration is disconcertingly easy to do. Oversimplification or reductionism is always possible, as well, even though a main purpose of case studies is to illustrate the multiple dimensions or rich textures of settings that policy documents and theories alone can render abstract. When the subject is emotionally charged by virtue of involving identity, agency, cultures, and cosmologies, fuller understanding can be hampered if certain views or concerns receive scant attention within the analysis. Many Africans know this well from having had generations' worth of experience being interviewed, photographed, scrutinized, and pronounced upon by outside behaviorists and theoreticians during the colonial era. The post-colonial era has arguably gone no better, since external financiers, politicians, and policy makers have dominated discussions about the continent's well being. Because of this history the present study has consciously emphasized its limitations as being the product of one external observer's perceptions of alternatives to a dominant development discourse at work in a particular African place at a particular moment in history.

But the story of Zambia's re-privatized copper mines, in fact the telling of alternative tales found there, would be incomplete without considering the experiences and insights of another group also present. This group consisted of the expatriate community, a comparatively small number of persons from European backgrounds who lived on the Copperbelt during the time of this study.[1] Expatriates in Zambia had their own valid sets of experiences. Awareness of these experiences could be particularly enriching for thinking about the relationships of policy making bodies and those expected to enact plans on the ground, for considerable tension existed there. Unlike the mine financiers and investors who were often headquartered in the northern hemisphere, mining expatriates were usually on location operational and technical experts whose responsibilities on behalf of the companies were to see that the actual work of production got done. Although their physical circumstances in the 2000s were considerably more comfortable than those of their counterparts in the early BSAC days,[2] many expatriates and the few Northern Rhodesians who remained in Zambia constituted members of a shrinking and largely unpopular fraternity. This contributed to discomfort of a different sort, for they frequently appeared caught in the middle between removed financiers who were preoccupied with production figures and Zambians who often maintained quite unflattering conceptions of outsiders across the board. This knowledge exacerbated an environment already fraught with unease and suspicion.

Shifting Attitudes Toward Mining

Mining had not always been viewed negatively, but by the mid-2000s shifts in the nature of what could be considered desirable white-collar work and the ecological movement's gathering strength had changed perceptions of the industry. In Western academic circles, interest in mining engineering had been on the wane since the 1980s and critical shortages of trained personnel were expected to hamper the industry well into the mid-twenty-first century. Some reliable old training institutions had closed down; others had amalgamated their curricula into more encompassing offerings on environmental matters. Bad publicity about the profession also affected the situation, especially when mining work was compared to careers in professions involving information technology. Production pressures, dangerous and dirty tasks conducted in far-flung environments seemed less rewarding than comfortable desk jobs in attractive, easier to reach areas.

> It's not looked upon as the good industry to be in these days. I mean there are all the "greenies" around attacking us, environmentalists also out there doing it, so it's become an industry you don't really want to be associated with anymore. It's also pretty hard graft. It's not the easiest way of earning money. So, you know, even guys who are coming out from mining universities these days are going into banking and analysis of mines and properties or the selling of mining equipment. Not that many coming through in what I'd call the old tradition of mining—going from shift boss and working their way up to whatever. And so it's getting tough all around.[3]

Compensating for these drawbacks within the privatized setting were lucrative salary and benefit packages that fueled the resentment of local staff and prompted occasional embarrassment even on the part of those who earned them. But, for many expatriates, the mining enterprise and specifically mining done in Zambia, was not simply about making money.

A number of expatriates had quite personal reasons for being in the profession and place where they were. The sensory environment of underground work was appealing to them, particularly the dynamism of an on-going exploratory project.

> When you get in a cage and the cage starts going down and you get the smells from the mine . . . it's a whole range of things that actually attack your senses and you actually feel very comfortable with yourself. And you go drive in town and you're on the surface and you smell lead fumes from cars and things like that. You get agitated and aggressive and—it's just the way it affects me. . . . You know, the greatest thing is when you get on the mine, you start it from nothing and it never ends. It's something you're continually building until the mine shuts. I mean, that mine is growing until the day it shuts. So it's so dynamic. It's not like a surface plant. You paint the floor today and you only have to paint it again next year. And it will still be the same. But here, every day you're moving. Conditions are changing, rock formations change. Different challenges.[4]

For some, there was also a gravitas to exploring regions never before seen by humans, as a geologist explained:

> You're dealing with Mother Nature. You're dealing with the rocks that—you know—what makes up the earth . . . Every time you go underground you see something different. It's never the same. Because you go into the new development areas where they've blasted over the last week or the last day or whatever since the last time you were in there and you're looking at something that hasn't been seen by man ever; if you think about it. And it's been there, in this particular case, for five or six hundred million years. . . . And I tell you it's quite sobering as well. When you think that you've never seen this before and you're here trying to do your best professionally for the company, obviously, and to a great extent for the country. You know we are also the policemen of the mine . . . this isn't a phrase that I thought up but we've been told in the past that we are the conscience of the mine.[5]

There was much above ground to admire as well. A noteworthy number of expatriates in Zambia's mines also had connections to the region: as white South Africans coming from that country's industry or as the children of parents who had served on the Copperbelt mines during colonial days. Europeans who had grown up playing in Northern Rhodesia's bush country and then returned during the re-privatization era remembered their early adventures with great fondness. Some seemed to have a visceral connection to the land and talked openly of Zambia as being their home.[6] For those with small children, the environment was considered a wholesome place to raise a family.

Accidents and Safety Disagreements

Despite a certain shared level of affection for their surroundings and interest in the mining endeavor, expatriates' opinions largely divided sharply from that of many Zambians over crucial work related questions. One of the most obvious areas of disagreement concerned whether or not the mines were more dangerous places under re-privatization than they had been under nationalization. A good argument could be made that conditions in the 2000s were actually *better* than before, although numerous changes in safety statistics reporting procedures that had taken place over the decades made definitive empirical comparisons impossible.[7] Injury and fatality rates remained unacceptably high by all counts, yet official ways of addressing these problems differed from popular understanding in both language and content.

At the most basic level, standard workplace terminology referred less to "accidents" than to "safety," a concept that corporate efforts promoted as being the shared concern of everyone in the industry. Mine safety involved taking personal responsibility for one's own life and the well-being of miners' fellow workers. Safety was also essentially a matter of the mind for, unless workers *thought* safely they could not expect to *be* safe. Indeed, when mishaps did occur, they could almost always be traced to something that demonstrated inappropriate or improper thought processes. Taking short cuts or circumventing standard procedures were classic examples, according to expatriates, of reasons why accidents took place. There were hundreds of points within the production process where workers could take short cuts or dispense with proper procedures. Insufficient barring down, or removal of hanging rocks, was a typical example as was failure to handle properly explosive materials. These materials were customarily cut in lengths that could burn for several minutes allowing time for the men who lit them to retreat before the charge detonated. Occasionally, however, workers would cut off a few minutes' worth of fuse. Then, if something delayed them on their retreat from the blasting area, they could be injured or killed. Failing to provide appropriate supervision and arriving at work inebriated were also typical reasons why miners might get hurt. If workers' minds were impaired by alcohol use or simply underused when called for, expatriates mused, then how could they not expect such mental attitudes to influence their being and acting?

Certainly a small number of annual fatalities could be attributed to things other than short cuts or circumvention of procedural rules. Suicides sometimes took place underground. Such incidents, however, were statistically negligible and their motives were usually discernable through post-accident investigations. Miners who died in suspicious circumstances might, for example, be suffering from relational or financial trouble in their personal lives. Murder or "score settling" also occasionally took place in the mines. But odd signs usually tipped off investigators to these sorts of deaths as well. Additionally, the numerical evidence of Zambia's higher than acceptable accident rate also had to be viewed in light of the fact that approximately 5,000 more men were on the job, working at deeper mining levels in the mid-2000s than ever before. More men at work inevitably translated into more accidents in the workplace.

So why were there so many accidents during the re-privatization era? Most expatriates attributed the problem to deferred maintenance during the nationalized period. Quite simply, Zambia's crown jewel industry had been allowed to degenerate for decades. Turning this around in order to make both the industry and the country economically viable necessitated ramping up production as quickly as possible. Additionally, since a global marketplace now determined policies and procedures in ways quite different from the colonial days, the mining companies had to conduct site renovations and proceed with production according to certain constraints.

Most notably, the privatized mines had to contract out work, saving investors' money by keeping a leaner workforce on their own books. Unfortunately, the advantage of saving money this way was offset by not being able to control key aspects of the subcontractors' practices. Chief among these was the issue of safety. From the standpoint of the mining companies, subcontractors under supplied their men with adequate training. There was sharp disagreement, however, on this point among expatriates who were in corporate management and those who owned contracting companies. Each tended to blame the other with the smaller specialized drilling or mechanical contractors finding the big companies culpable when the overall conditions of the mines were bad. Zambian law, for example, specified that work should not be done when underground temperatures went over a certain point. But some expatriate subcontractors claimed that this regulation was regularly flouted.

Blame assignation aside, fending off safety related problems absorbed a considerable amount of corporate energy and the mining companies' expatriate managerial staff members were often as deeply involved in accident prevention efforts as were their Zambian managerial colleagues. Standard corporate protocol called for the distribution of physical items such as safety gear (boots, hard hats, and the like); training sessions in proper techniques for using hazardous materials or potentially dangerous processes; education and awareness building through the use of signs, slogans, informative talks at the beginning of shifts; and so on. The availability of such resources did not, however, signify any more uniform accord among managerial layers within the mining companies than among the companies and subcontractors. Expatriate safety managers could feel production pressures intensely, sometimes sparring with upper level executives over conflicting priorities.

When accidents did occur, mining managers (both expatriate and Zambian) tried to use such incidents' aftermath as impetus for improving the work environment thereafter. One strategy—dreaded by the workforce and occasionally argued about among managers—was to discipline or punish those determined to have been responsible for causing an accident. Deciding whom to discipline, however, could be a complex effort based on imperfect and often frustrating investigations following mishaps. These official endeavors could be protracted affairs lasting for months and typically involving the workforce, management, and government's Mines Safety Department (MSD). Officially, such investigations were largely "fact finding" procedures that managers hoped would reveal both why an incident had occurred and suggest preventive lessons for the future. Since workers viewed these undertakings considerably differently, expatriate managers often encountered antagonism and resistance with workers lying or maintaining pained silences when queried.[8]

> [We] start off with the injured guy and find out what happened. Ask the witnesses what happened. Very often you might get five people with five different

> stories so you have to battle through the injured, and the witnesses, and the first line supervisor who would have been there. It's sometimes a battle to get the real story out of them. . . . Obviously, if we find the guy has done something blatantly wrong there's a big chance then we will have a disciplinary hearing later. So, obviously they are worried about that . . . but 90% of the time they are worried about that for nothing. We really just need to get the facts so that we can put something in place or change something if we need to. Very often it's actually the supervisor who's more in trouble than the injured.[9]

Unfortunately, expatriates in the industry generally felt they got little support from the government that had authorized the privatized mining arrangements. In fact, they often experienced representatives from the MSD and individual politicians as tending toward the obstructionist, blaming rather than helping the companies, acting arbitrarily and imperiously instead of collegially. It was not uncommon for the Zambian government to make public notice of fines, to threaten withdrawal of corporate licenses, and to punish, by suspension, particular managerial staff members. Following the cage crash in Mufulira, a group of government officials even attempted to go underground to view the accident scene, even though company officials insisted that this violated international mining regulations.

Support from mine financiers was also generally in short supply when accidents occurred. Although the role of investors in relation to mine safety and accident prevention was usually obscured by public clashes among unions, government officials, and on the ground mine company personnel, a few expatriates who ventured to comment on the subject thought that the investors generally gave no indication of long-term interests in Zambia. Such an attitude naturally affected how the companies approached general operations and safety concerns. Even company executives who defended their safety strategies occasionally noted the awkward communications vacuum that seemed to develop between the local mine and foreign corporate headquarters when accidents took place.[10]

Narrative in Expatriate Experience

The idea that, according to Zambians, supernatural forces could incite accidents was generally known but almost universally dismissed within the expatriate community. Certainly, witchcraft was understood to be part of Africa's traditional past and "superstitious" miners could be found in Zambia and throughout southern Africa. In fact, superstition appeared to be an aspect of mining culture worldwide.[11] According to expatriates, though, the power of supernatural experience in Zambia derived from the mental or emotional hold that it took on people, not from any real potency in the objects used or in unseen forces. What made witchcraft *work* was simply whether or not a person *believed* in it.

Similarly, knowledge of traditional stories about the mines and supernatural guardians of creation remained almost completely outside the realm of expatriate experience. A few non-Zambians had heard of old African smelting practices but seemed most interested in them from a technical, rather than cultural, point of view. Fewer expatriates were cognizant of the connection between mines and snakes. Nor did there appear to be much interest among the majority of expatriates in learning about such alternate narratives. This lack of curiosity about elements of Zambian cultural and religious experience was, at least in part, logically attributable to the fact that the expatriate population consisted mostly of businesspeople and technicians rather than academics and cultural researchers. Disparities in intellectual or cognitive overlap did not mean, however, that Copperbelt area expatriates lacked their own narratives or philosophical views on the privatization matter. In fact, a broadly common narrative appeared in the ways in which most expatriates discussed the privatization process and its context. The account could be summarized roughly in the following manner.

Zambia's economy was historically based upon copper mining. If the nation wanted the mines run properly, there were certain jobs and essential skills required. Mines always need personnel trained in metallurgy, engineering, and smelting. But the country had not produced these sorts of workers in any significant number since before the era of globalization. So, from the time Zambians took over in the 1960s until they relinquished control in the early 2000s, the mines had a chronic skilled labor shortage.

Compounding this issue, nationalization and Zambianization then put people in positions of responsibility who had neither the skills nor proper personal attitudes to run the mines. This sort of inexperience was a main reason why precious mining income and valued personnel talent were diverted from the mines in order to support rapid and extensive national development. In the meanwhile, who knew how much money meant for the national welfare had gone into personal accounts? It was also obvious that no reinvestment had gone into the mines themselves under Zambian leadership. And there was simply no excuse for blaming the mines' collapse on sagging world copper prices; the economic data and statistics clearly showed that this could not have been the case.

So, the real reason ZCCM failed was simply that Zambians had not been good managers. Unfortunately, it did not appear in the 2000s as if they had learned from the mistakes of the previous decades. In the era of re-privatization personal interests continued to trump organizational loyalty; long-term vision or planning remained virtually non-existent; and a willingness to accept personal responsibility for their work or to hold fellow employees to account just seemed absent from the Zambian work ethic. These were all matters of choice, however. Poor initiative, for example, was not due to innate indolence. Zambians often displayed little initiative because years of living off the state's support systems and the patronage of foreign assistance had accustomed them to wait for others to fix their problems. Poor ability was not an issue pertaining to the number of brain cells a person had. Zambians often performed poorly because they seemed unwilling to use the intelligence they possessed!

In the end, such attitudes and actions ultimately cost Zambia most dearly. If giving job security to one's relatives and friends was more important than finding competent people to fill top corporate positions; if remaining loyal to tradi-

tional group norms was more valuable than encouraging individual excellence how could Zambians ever expect their companies to be profitable? No wonder the most gifted and talented of the country's citizens moved into private industry (where values of efficiency and accountability predominated) or left the country altogether. Unfortunately, this "brain drain" left Zambia's government, civil society, and most of its labor force with disproportionately higher numbers of individuals lacking the education, training, and skills to run the country.

Given such a situation, Zambia's mining industry would always require a small contingent of Western managers. At least for the foreseeable future, expatriates would be needed to fill the companies' most technical and responsible jobs; there simply were not enough Zambians around who could do them. Additionally, adopting a Western cultural value system that preferred organizational to familial loyalty would be necessary to help counter corruption and nepotism. Western acumen for future-oriented planning and systematic thinking would help ensure that Zambia reinvested in the mines to keep them running. So, if local people resented the expatriates' presence, maybe they were simply angry and jealous (even embarrassed?) that the mines were once again producing under privatization when they had not done so as nationalized companies.

Given this history and from the vantage point of expatriates, unconstructive criticism that the Zambian public and government routinely piled on the industry was both ludicrous and exasperating. During the 2005 spate of accidents, for instance, Zambian government officials had been scathing in their condemnation of particular mines' safety programs yet seemed conveniently to overlook similar trouble in other industries and the inadequacies of their own agencies. Valuable company time and effort went into responding to this sort of unhelpful opposition and it personally exhausted mine management as well.

Non-Zambians also felt unfairly judged and criticized by Zambians on any number of levels with the issue of corruption being particularly irksome. Virtually everyone agreed that corruption had become comparatively well entrenched during the ZCCM days. But the extent to which corruption carried over into the new companies was a problem of varying degrees depending on one's perspective. Despite the Zambian government's and media's publicly expressed anti-corruption sentiments and similarly well publicized anti-corruption efforts of Zambia's international donors, the amount of genuine, collective interest in opposing corruption was questionable. Mine managers knew this first hand because some had experiences of their own staff's involvement. Their attempts to disrupt such practices as kickbacks and unorthodox contract letting within the companies they ran could trigger intense private resentment among some Zambians.[12] No matter what they did then—turn dilapidated mines into productive ones; convert dysfunctional work practices into functional—they generally felt themselves surrounded by an aura of suspicion and distrust.

Philosophical and Theological Puzzlements

The predominantly mechanistic or operational tone of a narrative that seemed commonly expressed by expatriates was often overtly linked by them to morals and ethics. In fact, quite a few felt frictions between their own ethical systems and those they saw being lived out by citizens of an officially declared "Christian nation." Often, expatriate expectations of how Zambians should be and act did not match what seemed to happen in practice. Outsiders noted, for example, an apparently loose relationship between word and deed in their Zambian colleagues' lives. The giving of false reports on the status of projects or bogus testimonies at accident investigations were commonplace examples of such discrepancies. To expatriates this was an ethical violation. How could people who expressed such overt religious piety so deliberately make false statements? How could Christians do what expatriates knew Zambians sometimes did in the workplace—such as flee the scene of underground injuries or otherwise not pitch in to help co-workers involved in accidents?[13] This sort of behavior appeared especially puzzling in light of Zambians' concerns about accident rates underground. To expatriates, such instances seemed clear cut examples of intentionally saying one thing and doing another, of not taking responsibility for individual actions.

From the strictly intellectual standpoint of Western philosophy such assessments seemed to be quite valid. It would be difficult to use the tools of logic and empiricism, however, to reconcile general perspectives of expatriates with many aspects of Zambian behavior. So it would seem that facets of alternate narratives glimpsed in preceding chapters here become particularly significant. Such glimpses potentially offer clues as to how policy makers and practitioners might assess not only the ethical systems of others but certain thinking patterns that have maintained an intellectual hold on the development industry itself. This is not to say that expatriates should have abandoned their own philosophical systems and their occasionally overtly expressed theological sentiments in favor of adopting something else. But, as the concluding chapter will attempt to point out, there may have been some culturally legitimate reasons all around for the moral conundrums expatriates experienced. Awareness of factors potentially affecting Zambian thought and action during the re-privatization era could be helpful knowledge for more effective policy making elsewhere in the future.

Notes

1. For the purposes of this study the term "expatriate" applies to persons of white, European extraction. See the discussion of research methods in chapter two. According to the mines' annual reports and human resources managers, the number of expatriates working in the industry at independence was about 16% of the total (approximately 7,300 whites within an overall number of 45,000 miners). *Annual Report of the Mines Department 1968*, 1. By the era of re-privatization, according to mining company officials, the number of expatriates working on the Copperbelt was estimated at 3%.

2. Expatriates who lived on the Copperbelt during the re-privatization era did not have the full array of amenities that had been available when the European population was much higher. In the first decade of the 2000s, there were several restaurants and private clubs, a few gymnasiums, and a modest cricket field or two available as well as some golf courses. But organized activities of broad appeal to the expatriate population such as theater and music performances were comparatively rare. Some expatriates commented on the lack of night life; satellite television and the Internet seemed to have taken the place of many previous forms of entertainment.

3. Executive, Copperbelt mining company, interview with author, 2004.

4. Safety manager, Copperbelt mining company, interview with author, 2004.

5. Geologist, Copperbelt mining company, interview with author, 2004.

6. Whites were generally not offered citizenship or passports, however, unless they renounced citizenship held in any other country. Many were unwilling to do this.

7. The country's mining industry yearbooks ceased publication of certain standard accident statistics in the 1980s. Statistics recorded for the 1990s are figured according to different criteria. See chapter five.

8. This was not merely a phenomenon of the re-privatized context. Fields speculates that colonial investigations of Watch Tower witch-cleansing episodes encountered court testimonies deliberately designed "to conceal facts, to shift responsibility, and perhaps to take advantage of the examiners' naiveté about witchcraft and the inner workings of the villages." *Revival and Rebellion*, 191.

9. Senior manager, Copperbelt mining company, interview with author, 2004.

10. For instance, in 2004, a senior executive commenting on a multiple fatality incident said he had sent three emails to the corporate headquarters that day and received no reply.

11. Expatriates who had worked on other continents remarked on this and a coal miner in Nova Scotia recounted some of his coal miner father's observations in this regard.

12. For example, when one mining company identified a group of mid-level employees engaged in corrupt procurement procedures and kickbacks, members of its expatriate staff found themselves pitted against Zambian nationals as the company attempted to stop the process.

13. Mine safety and managerial personnel reported that such instances did occur, sometimes resulting in needless fatalities.

♀

Chapter Nine

Conclusion

This work has considered how a particular African community appeared to experience the privatization of its main industry, a strategy that has been common practice in large-scale development interventions for several decades. The study has, however, considered its subject not principally from an economic or political viewpoint but from a perspective that gives voice to cultural and religious understandings. Findings gathered from in-depth investigation conducted over a period of years suggest that the general populace maintained some viewpoints about the privatization process that differed radically from what mainstream development theories on the subject would suggest.

Recapitulation

The exercise of selling Zambia's distressed mining industry and other parastatals, or nationalized companies, to foreign investors was hailed by policy makers as integral to the country's well-being. But things looked very different to ordinary citizens who resided on the Copperbelt. Even several years into the process, as the mines' technical capacities were being rehabilitated and copper production greatly increased, most Zambians were still not seeing positive changes in their lives. In theory, it is reasonable to caution that the benefits of such huge structural changes to a country's economy can take time to affect the general population and that free market strategies produce uneven—but ultimately broadly beneficial—results. The difficulty for Zambian miners and their families, though, was not just that they did not see themselves as benefiting from privatization yet. It was also that the public, or official, ways of dealing with the privatization process did not fully account for many of the problems they saw plaguing their society. Yet some of these problems seemed directly related to what was happening with the mines.

Indeed, for centuries Zambians had conceived of copper and of the underground quite differently from the ways that professional development discourse portrayed these things. Although Africans had engaged in mining, smelting, and copper trading long before Europeans arrived, they seemed to value the mining environment not as a commodity or resource but as something primarily of religious and cultural significance. In fact, the mines and metals were integrally related to mysterious forces that operated in both spiritual and physical worlds.

Almost a century before the re-privatization effort described in this study, the mines had been industrially developed by European outsiders and much of the disagreement in the 2000s over the role and function of the copper mines may be traced to this circumstance. From political independence in 1964 up to the late 1990s the mines operated as the premiere company within an array of nationalized companies. Zambians had taken over the mines' operations shortly after the country's independence and copper production remained an important function. But it was not generally considered the only thing that the mines were about. The mining companies were responsible for a host of other interests as well, especially the provision and maintenance of social infrastructure. Compiled statistical evidence of this period confirms that the country's main foreign exchange earning industry and the government itself fell increasingly into economic and managerial disarray. Yet there did not seem to be a broad consensus among the population as to why the mines should have been sold as they were under a program designed and implemented by the World Bank. In fact, many ordinary citizens actually attributed the cumulative economic and social ills of the time to the mines' sales and all that the new arrangement of Foreign Direct Investment (FDI) entailed.

Zambians especially worried about the number of mining accidents that were occurring in the post-privatized environment. In their experience, these incidents had increased dramatically for a variety of reasons, FDI and the renewed presence of expatriates being among the most obvious precipitating problems. Both the World Bank and the Zambian government had said that FDI would ultimately create a wealthier society. But in the early to mid-2000s private ownership seemed not to be offering widespread opportunities for competitive entrepreneurship but perilous life-death threats against the populace.

The situation could be explained publicly in operational, or technical, language of the type generally seen in development policy making. Consequently, an increased accident rate could be attributed to attitudes and practices of the private companies and this sort of talk did go on among Zambian Copperbelt residents. Many maintained, for instance, that companies that valued profit over people were bound to be careless about the industry's inherent hazards. A predominant view among Zambians also seemed to be that so long as the companies met production targets, the investors and managers did not appear particularly concerned if miners were injured or killed.

Such a mechanistic way of explaining the situation,[1] however, was also generally not adequate for Zambians in so far as mining accidents or overall misfortunes were concerned. In fact, it appeared that the reason why any unfortunate event happened could not be discerned by rational, theoretical, or operational analyses alone. Accounting for accidents and mishaps within the everyday experience of Zambian Copperbelt dwellers frequently involved a kaleidoscopically shifting array of interactions among individuals, corporations, immaterial forces, and material interests. Specific moments of crisis made this acutely obvious, prompting Zambians to search for patterns of relationships or various series of conditions that could explain them. To complicate matters, however, relationships or patterns that Zambians saw were not necessarily or immediately obvious to outside observers. In fact, many of the thoughtful connections that Zambians could find among circumstances were viewed as totally unrelated by non-Zambians such as expatriates and outside consultants associated with the re-privatization process.

For Zambians living through the privatization process empirical facts were not the only types of relevant information they had to consider. They actually appeared to make sense of their situations based on a cosmological understanding that did not parallel the understandings embedded in official, or public, arenas where policy was made and discussed. Most notably, unlimited by time or space and unleashed at least in part by behavior associated with the new economic regime, supernatural or spiritual forces played a powerful role in what Zambians perceived to be happening.

In the average Zambian miner's experience of the country's shift from socialist to capitalist economic principles, witchcraft could be employed against those who seemed to become too successful; blood sacrifice rituals could be conducted in efforts to gain wealth and power for an individual or family. Even the simple day-to-day hardships of anonymous families could be cosmically inter-related with larger societal matters. The series of additional misfortunes that Joseph's family experienced after his death in the BGRIMM explosion, for example, were not just unfortunately timed random events. They were somehow of a piece with the overwhelming loss of his person—little residual ripple effects—that could be expected in such cases and simply had to be endured.

Supernatural involvement in everyday Copperbelt events during the re-privatization era could also be found on a more widespread scale. This was evident, for instance, when some national media contributors and private citizens pondered the 1970 Mufulira Disaster and 1993 Gabon air crash in relation to the 2005 spate of accidents or when speculation circulated that President Mwanawasa was sacrificing miners for his own power needs.

Analysis

A cautionary word should be offered at this point about the intended aim of analyzing this case for it can be presumptuous to speculate about what others think, feel, and value. Africans throughout the continent have been particularly presumed upon in this way by outsiders. So, it should be understood that the comments to follow are not offered as definitive statements *about Zambians*. Rather, they are meant principally to provoke self-reflection on the part of well intentioned people, especially Americans, who are interested in the common welfare and who may, themselves, have an active life of faith or philosophical grounding. For, as this study has maintained from the outset, discrepancies in understanding that occur across cultures have theo-ethical ramifications. Analytical comments are also offered in the hope of identifying elements of Zambians' viewpoints that could be employed to temper the intellectual framework underlying much current interventionist policy, whether such policy comes from secular or faith-based sources.

Zambians appeared to participate in an experiential realm that ranged beyond the intellectual parameters shaping much of the re-privatization effort's policies and practices. This state of affairs offers at least two important clues concerning how they made sense of their everyday worlds and especially of the re-privatization process. First, the boundaries of what could be considered legitimate contributions to general knowledge or reasoning processes were expanded.[2] That is, there did not have to be a logically obvious link (as outsiders might see things) between circumstances in order for Zambians to perceive an inter-relatedness. Rationalism and empiricism remained valid forms of knowledge or information seeking in the Copperbelt context, in fact, they could be creatively used as the Coke-Fanta experiment at Kansanshi illustrated. But rationalism and empiricism were not considered the only sources of legitimate knowledge for Copperbelt Zambians. Intuitive knowledge, rumor, story, and information gained from experiences of the supernatural realm were also considered valid.

Second, in everyday life and work there were simply more factors to be considered and less certitude about the dividing line between them than official development policy making could accommodate. It appeared very clear, for instance, that the supernatural beings, or forces, that miners talked about were not vague representations of an inherited cultural knowledge base in the way that many Westerners might share a Grimm's fairy tale background. These were real entities posing practical problems, not just theoretical challenges for Zambians. For instance, the magical snake that kept appearing in the church near Luanshya did not cause difficulties just at the psychological or theological level. What seemed to worry Zambian Christians was not that the snake's presence might mean something significant about their own mental well-being or religious understanding. Rather, the most immediate dilemma the snake's presence posed for them appeared to be that the thing kept vanishing before anyone could pick it up and take it away for disposal. Similarly, classic Western Christian questions concerning the ethics or morals of having an affair were not necessarily the most pressing issue in townships where "private sectors" were being conducted. The most immediate issue seemed to be the affair's timing since that might endanger a cuckolded spouse while on the job. Further, very real, but unseen, spiritual forces could imbue material objects with power allowing them to defy the normal ways that matter was generally understood to act. Actions akin to what quantum physicists might conceive in theory were commonplace elements in Zambians' experience of the supernatural. Peoples' behavior could change unpredictably after coming into contact with magical substances or *muti*. Human beings could transform themselves into other creatures or they could even fly on their own.

Living in such a realm of experience seemed to affect Zambian self-understanding in numerous ways. On the one hand, the existence of diverse, potent, and unpredictable spiritual forces acting within the material world presented people with far less autonomy, or personal agency, and more vulnerability than outside observers might appreciate.[3] From the Zambian standpoint, how could a person innocently navigate through an environment in which one's neighbor might change corporal substance or secretly secure *muti* for malicious use? How could one ever be sure of reality when a host of principalities and powers routinely made themselves felt, prompted by unknown motives? Such an environment might reasonably persuade Zambians to be fairly tentative about their surroundings. But on the other hand, provided they were willing to engage in risky supernatural encounters, people actually seemed to have more options for what they could do with their own physical bodies and with the objects of daily life than might be initially assumed. Stupendous wealth could result, for example, not from engaging in arduous and unglamorous work, but from having secured proper human body parts for sacrifices or from being magically "cooked" to achieve superhuman abilities.

Patterns of thinking observed among the general Copperbelt populace of the 2000s posed distinct challenges for views associated with official policy making and processes of the re-privatization effort. Consequently, it can be said that the experiences of Zambians in this situation offer profound alternative viewpoints, or tell a radically different story, from what this work has termed a dominant development narrative. In the West an appreciation for the sort of intuitive, supernatural experiences described in this case study largely parted ways centuries ago with the intellectual pursuits that have produced science and the modern age of development.[4] Indeed, as was mentioned at the outset of this work and as is increasingly coming under scrutiny by development studies experts, a commonplace theory undergirding such external interventions has, in line with the secularization thesis, assumed that "backward" or "superstitious" beliefs about unseen spiritual forces will dissipate once sufficiently disproven by scientific-rationalist methods of inquiry. The evidence that this is actually not the case—certainly not on the Copperbelt of the early twenty-first century—highlights obstacles and opportunities for people who want to engage across cultures on behalf of the common good.[5]

A few words should be interjected here about the theological ramifications of this situation. Volumes have been written about the meaning and significance of syncretism: what constitutes syncretism; what constitutes "genuine" Christianity. Certainly, theological investigations into authenticity or legitimacy of practice and understanding can be important and worthwhile. And, if one wanted to concentrate on this aspect of the Copperbelt case, there would be ample fodder for doing so since there often appeared a blended theological understanding within the experiences of Copperbelt Zambians. Sometimes, for instance, it appeared more appropriate to talk about circumstances of misfortune in terms of Satanic practices and beliefs—as people did when analyzing the 2005 spate of accidents and supernatural experiences taking place in urban settings—than to use other words such as "witchcraft" and "wizardry." Sometimes it could be legitimate to complain on the basis of a Christian identity about pluralistic influences, as happened when Indian owners of the Luanshya mine put Hindu idols underground. Simultaneously, however, it appeared acceptable not to worry about the legitimacy of a Christian community's authenticity if a supernatural snake showed up inside the church. This evidently blended theological understanding at times both puzzled and frustrated expatriate outsiders. When they could not make sense of how Zambians seemed to connect beliefs and practices, words and deeds, the authenticity of Zambians' Christianity could be called into question.[6]

But, beyond the issue of how Zambians regarded their own apparent multiple religious belonging, the Copperbelt case suggests another theologically significant point that frequently escapes outside notice even though it is crucial for practical situations of cross-cultural engagement. That is this: if human life is understood as sacred and valuable, then it is incumbent upon us to take seriously how people in one culture identify and define problems even if what results defies adequate interpretation according to our own worldviews. If, for instance, Christians at the Luanshya church appeared to have no real anxiety about the legitimacy of their faith in light of supernatural snake manifestations, perhaps the theological energies of outside observers should go not so much toward pondering what it meant to be a Christian in that circumstance but to what Luanshya residents saw as the meaning of the snake's presence. To take another example, thinking about how Copperbelt residents analyzed connections among misfortunes, blood sacrifices, Satanic activity, power and wealth accumulation can lead to fascinating theoretical expositions. But focusing on how such supernatural activities might or might not accord with Western Christian understanding can allow something else of critical importance to elude outside observers, namely that Zambians were experiencing misfortunes they attempted to explain.

The point is that development interventions involve concrete actions with palpable effects and while those who experience such interventions may not talk about them using syntax that policy making has typically been able to accommodate, there is an ethical imperative for policy makers and practitioners to listen. The cosmological understandings of Copperbelt Zambians experiencing privatization were such that genuinely effective development policy in that arena should have, at some substantive level, engaged philosophically, culturally and theologically with them. Arguing, as some might, that an economic institution such as the World Bank should not be expected to take on theological matters would largely miss the point being made here. On one level, as Deneulin has observed, since there are many contexts in which religion cannot be separated from development, it can reasonably be said that Bank officials actually should have done more in this regard. Post-privatization admissions of inadequate attention to cultural matters suggest that at least some officials within the Bretton Woods institutional system recognized this.[7]

But on a deeper level and specifically in reference to this work's target audience, the point is that people who are overtly theologically and philosophically inclined, people of faith, and especially American Christians engaged in international outreach, have both ethical imperatives and excellent rationales for refusing to let the intellectual terms of their engagement be defined by economists and politicians. What would this mean for a place like the Copperbelt generally? Recognizing that Copperbelt residents were religious and involving their faith communities in development efforts would certainly be better than acting as if religion and faith did not exist. But it would still be an inadequate approach if the objective were to work in cross-cultural partnership for the common welfare. Needed in that circumstance would be deep, self-reflective efforts to attempt understanding not simply of another culture's worldview but of what that worldview implies for living in the everyday. Ultimately, such endeavors should qualitatively affect at least some international, cross-cultural engagements.[8]

Toward such an objective, the Zambian case presents information useful for revisiting certain assumptions that appear inherent in much development policy making and implementation. Three examples that are highly relevant to theo-ethical concerns include conceiving of the world in interconnected terms rather than in discrete terms; broadening and deepening discussions of relationships between identity and power (or being and acting); and investigating how different understandings of material and non-material realities affect understandings of cause and effect.

A struggle over different manners of thinking—whether interconnected, discrete, or something else—was evident throughout the case, most tangibly in Luanshya's painful experiences with the Roan Antelope mine. When new Indian owners buried Hindu idols underground, the local community took great offense. Zambian miners responded by urinating on the idols and it was the idols' presence that prompted predictions of Binani's short-lived ownership. Analyses of this situation informed principally by secular development theories attributed the mines' problems to corrupt politics and incompetent management. Although this study found no official development agency references to the Binani idol controversy, had pronouncements about this situation been made by such organizations, the miners' actions would likely have been defined simply as rude objections and the idols themselves as mere symbolic objects. Similarly, the Zambians' protest would probably have been explained as illustrating consistency with what workers might do in any heated labor-management dispute.

It could seem that concentrating on the idols was simply an excuse used by Zambians who were ill-equipped to handle work place conflicts. If they were the Christians they claimed to be, why not just ignore the statues and fight more important battles for their own welfare? Furthermore, what did it really matter if the end result—Binani's departure—was the same as that predicted by industry analysts and policy experts? In other words, perhaps all the talk of underground mayhem due to idols and the institution of Hindu religious holidays was simply the way Zambians dealt with poor management and unpleasant intercultural frictions if they were not sophisticated enough to use business and managerial discourse.

But neutralizing a sacred object's power is different from protesting its presence. Predicting the company's demise because of provocative physical items set out by management is not the same as predicting corporate failure due to management's operational decision making ability. Most notably for Zambians, desecration of the mines seemed not just another example of cross-cultural miscommunication but a key aspect of the company's failure.

In fact, as this work has made clear, Luanshya miners had a long history of dealing forthrightly in labor-management disputes over wages and matters pertaining to compound life. During the Binani controversy, they were able to dissect quite incisively not only the company's managerial faults but also government policies that had led to the sale in the first place. These conversations went on simultaneously with the uproar over the idols and continued for quite some time after the mine had closed down. So, Zambians seemed quite capable of distinguishing between managerial or business sorts of problems and other, interrelated sets of problems. There was no doubt that Zambians found fault with Binani management. But it could reasonably be argued that they saw the mine spirits' displeasure at treatment of the underground as the *ultimate* signal for why Binani's tenure was short-lived.

Worker fixation on the Hindu idols also illustrates an important point concerning the location and exercise of power as these relate to being and acting. From one vantage point, the most potent forces appeared to be those of the physical mine and the Hindu idols. Consequently, the mine spirits disapproved of being challenged by foreign objects and interests. From another vantage point, power seemed concentrated in the managerial and labor processes of modern industrial mining. While other perspectives might well reveal additional power focal points, the issue here is that where power is understood to reside can make a monumental difference in practical, cross-cultural work settings. Understanding ultimate power as residing in physical places overseen by spiritual forces is very different from conceiving of ultimate power as vested in processes devised by the creative will of individuals and groups. When disparate sources of power engage with each other—especially when they are understood within different cosmologies—complications of incalculable proportions can result.

This was especially evident in the Copperbelt debate about accidents and safety. Zambian miners who used the industrial companies' safety equipment and procedures along with their own measures such as taking shells and written prayers underground may have done so because all these forms of material objects offered protection. The types of things they protected against, however, were different: an apron or set of goggles might protect a miner against a spray of hot metal from the smelter. But there was no guarantee that such equipment would repel whatever motivated the spray in the first place. Something more was needed for that. If a shell or bit of wood could protect against an incident's provocation, was it not then more powerful than the sturdiest gear provided by the companies? Further, since people could really never know what might provoke a haunting spirit attack or jealous witchcraft episode, how fair was it to say that safety was mostly a mental attitude?

Alternative interpretations such as these deserve to be taken seriously by Western people of faith, analysts, and policy makers. Doing so in the Copperbelt case would not have to mean advocating more relaxed corporate safety measures or insinuating that mining companies were wrong for insisting that workers adhere to proper procedures. Rather, the Copperbelt case suggests serious consideration because of the possibility that Zambians held a qualitatively different understanding about the roles that equipment and miners themselves could play in creating a safer working environment. Workers may have put less emphasis on their own powers or abilities to affect the situation and more on a range of *objects* to do something because of how they thought about power and how they viewed the forces of the material world. Perhaps they simply assigned more intrinsic power to certain material objects than would others and this assignation affected how Zambian workers sometimes acted.[9]

A perspective that found intrinsic power in the material world further appeared to affect many Zambians' conceptions of immaterial realities. This was particularly obvious in language matters. Rather than simply representing opportunities for semantic quibbling, Copperbelt linguistic concerns posed provocative questions regarding assumed relationships between word and deed. What if, for instance, different understandings of what words, less by their meanings and more by their utterances, could *do* affected workers' actions on the job? The careful observance of proper verbal goodbyes and greetings that bracketed miners' workdays might have arisen from an understanding that the spoken words themselves bore on mine safety. Insistences by some Zambians that the days of ZCCM's "accident prevention officers" had been safer than those under the private company "safety officers" might have been expressions of more than resentment at institutional change. They might also have been saying that an "accident prevention officer" was simply better empowered to protect than was a "safety officer."

Intriguing possibilities for relationships between word and deed might have mattered especially to post-incident investigations. Miners' lying about the causes of underground fatalities could be understood as nervous attempts to protect their jobs against potentially punitive investigative processes. But if words *on their own* could incite further misfortune, perhaps lies did something else as well. Perhaps they helped constitute invisible cloaks of protection that would give haunting spirits no reason to suspect ill will on the part of colleagues who remained alive. From this vantage point then, rather than inhibiting corporate safety procedures, a lie could actually *contribute* to accident prevention. Likewise, expatriate complaints about working within a general arena of distrust and suspicion may have indicated their inadvertent encounter with a cosmological system in which words had a qualitatively different sort of efficacy. If, in Zambian experience, words had innate potency and force, there might have been a sense in which at least some Zambians felt that *saying something* really would *make it so*. Saying a project was completed might, for instance, really help progress along even as it kept peace with the bosses in the interim.

The above points should not be viewed as defending workers' "factual" misstatements or as asserting that managers were wrong to expect a certain sort of honesty and action from the workforce. Neither does this sort of speculative interpretation represent an attempt to overlay a romanticized vision of Zambians as performing "incantations" on the job or to constitute a patronizing interpretation of them as being incapable of thinking in modern terms. Rather, the means of comprehending relationships between word and deed suggested here may illuminate something qualitative about alternate points of view. In this case, what might appear as an ethical lapse (lying) from one vantage point could be understood as a pragmatic action (preserving one's job and keeping a general peace) when viewed differently.

Thus far, the evidence suggests that many Zambians inhabited an experiential environment in which the boundaries between material and immaterial realities could easily shift while a variety of powers could routinely complicate efforts to understand the causes or the origins of events. In short, the world simply behaved in ways that official development policies and procedures could not address. What, then, might such a viewpoint say about how dominant narratives depict material reality?

In "developed" society, material infrastructure certainly holds great meaning. But the material usually has to do something or sit within a set of larger objectives that may, at least in part, be immaterial realities. For example, cell phones, while desirable in themselves, are only as good as an invisible but functioning network. Programs and projects are nice to talk about but are only as real as what they accomplish, not as simply described on paper. To a great extent, then, the equipment and activities involved in getting the job done are typically depicted within development discourse as ancillary to the central issue.

From the Zambian perspective, however, possession and use of development's perquisites (TV sets, cell phones, and vehicles; meetings, workshops, and travel) may have been so important at least in part because the items and activities themselves were understood in a more substantive fashion. Zambians may well have wished to enjoy even a fraction of the luxuries they saw used by expatriates and on imported television shows because the same attraction that, for instance, many Americans feel for development's amenities affected Zambians, too. But what if many ordinary Zambians somehow also regarded the goods and the get-togethers differently than policy experts might expect? What if, rather than being vehicles for getting to an objective called a "developed" state, the items associated with the processes, such as computers and cell phones, along with activities such as meetings and workshops were viewed as the desirable end result?

The distinction here may be fine. But it is important and perhaps most clearly understood by returning to what some development studies experts have already said. To reiterate from the opening chapter when, for example, Amartya Sen argues that a properly holistic understanding of "development" sees it as constitutive of the *processes* that lead to or produce freedom,[10] he is appropriately expanding the intellectual basis upon which much development policy is made. Chinweizu's observation that goods and bureaucratic processes can *themselves* be seen as development's necessary ingredients[11] also expands the intellectual basis of development policy making. So, just as it is wise to understand "development" as abstract processes leading toward self-determination it is important to be aware that "development" may sometimes be understood as primarily about concrete, physical elements themselves. This latter point should not be viewed simply as a double back to chapter one's discussion of materialist development for there remains important nuance here. Development as understood in materialist terms could still be said to rest upon an assumption that the physical items have to work. A built school would be expected to house activities related to the education of children; a built hospital would function similarly with sick patients. At least one other way of looking at the situation exists, however. This way speculates that there may be cases in which the connection between the material object of development and its function according to dominant theory is less important than the existence and possession of the item itself.

Understanding the relationship between material and immaterial realms in this way would help explain why the privatization process evoked fierce disagreements over what the mines were; how they were expected to operate; and what the entire industry was supposed to accomplish. According to the narrative associated with public development discourse, once Zambians had taken over mining operations they had not concentrated on core business. Instead, they had turned a large and profitable mining company into something of a social services and development organization. The unfortunate results of that state of affairs were mismanagement and production loss which ultimately became the reasons for re-privatization. This line of reasoning was defensible by statistical data and some historical accounts, so why did Zambians seem not to understand what they needed to do to ensure future success?

Perhaps the answer is in part that Zambians had an inherently different understanding of what the physical environment *meant*. Perhaps for some Zambians the experience included an understanding (or hope?) that simply having the accoutrements of the mining industry would ensure their personal well-being. Might it be possible that, for many Zambians, the physical attributes of the mines were actually more *real* than the intangible thought processes and ideas needed to keep the mines operating? For present day Zambians, could going down into the earth to extract copper constitute a greater reality than would consideration of graphs and statistics for future explorations?

This set of possibilities makes not only Luanshya's story but details such as the ZCCM/LSE ore classification systems issue intriguing. Could it be said that, in addition to the technical training and abilities that Zambian mining managers did possess, something suggested that mining operations were ultimately more "real" when conceived in terms of physical infrastructure rather than as intangible commitments and numerical projections? Could a similar understanding have, at least in part, precipitated the lack of exploration for future deposits that characterized the ZCCM years? Certainly, the shallow pool of competent managers and multiple demands being made on government competed for attention with the business of mining. But could there have been a sense in which ore exploration was accorded less importance than other development activities because the ore presently at hand was simply so *real* that future planning was not deemed urgent?

Speculating about substantive differences in understanding of material and immaterial realities should be done with an understanding that most Zambians were at an historic disadvantage in gaining knowledge about how Westerners comprehended the world. For the most part, Zambian miners' experiences consisted of only marginal exposure to expatriate managerial practices gained primarily visually and at a distance.[12] This exposure would have included memories and stories of how earlier era mining executives and their families had enjoyed comfortable houses, good clubs, and so on. Such memories and stories generally would not have been accompanied by close access to Westerners' ways of thinking, however. So, even though Zambians who took over mining operations in the nationalized era might have heard talk and seen photos of how things used to be, their opportunities for learning in detail about the systems and processes needed to create and maintain a "developed" society would likely have been minimal.

This possibility can be helpful for analyzing disputes over what development is supposed to be about both now and in the future. If the possession of certain items—rather than management of processes to maintain and enhance them—is understood as the essential ingredient for having the sort of well-being that development is supposed to bring, then Zambians' emphasis on acquiring goods rather than on constructing systems or making plans becomes more comprehensible behavior to outside observers. In other words, the Copperbelt case suggests that in cross-cultural work place settings, while some employees might concentrate on acquiring facts and adhering to certain intangible beliefs or principles, others might concentrate on acquiring physical items and performing certain routines. This would not mean that some were really working while others were simply going through the motions. All might well understand themselves to be properly engaged in work necessitated by principles of development. The difference would depend on their interpretations of what *work* really was. This, in turn, would hinge, at least in part, on how everyone conceived of material and non-material elements involved. All of this would ultimately affect various understandings of "development" itself.

It is only a step from debating what the mines and miners should *do* in such situations to contemplating who such workers should *be*. Many Zambians seemed to understand that they were capable and skilled yet subject to a variety of forces and powers including the detached disinterest of non-Zambians. Most intriguingly, it appeared that some Zambians who were educated, Westernized mining executives did not feel as comfortable with their own identities as outsiders might assume. To put it forthrightly, even when occupying jobs of considerable responsibility and importance, elite members of Zambian society could exhibit signs of what Westerners might term low self-esteem. Certainly, some of this situation could be attributed to the lingering effects of colonization.[13]

But there may have been more than political and economic reasons for what outsiders could see. If, for instance, Zambians understood themselves religiously and culturally as being born into a powerful and mysterious cosmos in which they were comparatively vulnerable, this would have shaped how they construed their identities. Historic experiences of colonialism along with subsequent events such as the Cold War and their country's acquisition of international debt would simply have corroborated an already innate sense that they had limited ability to control their own destinies. If such were the experiences of Zambians, it would be unreasonable to expect that offering education and work modeled along Western lines could sufficiently overcome general feelings of lack of self confidence or personal power. Despite their outward appearances and innate abilities, many Zambians may have been in quite high level jobs without feeling a sense of autonomy and agency comparable to that perceived in their European or American counterparts. Simultaneously, their expatriate counterparts may have assumed that their Zambian colleagues did feel a sense of ability, agency, and confidence equal to their own.

While our limited access to how others perceive reality means that discussions of personal autonomy, agency, and power are quite speculative, they are still important for policy makers and practitioners to consider. Learning from the Copperbelt case it could be said that a perspective that understands humanity as autonomous and free from elements of fear that witchcraft and supernatural experiences can engender: a perspective that concentrates on mastering (more so than reverencing) its environment is in a stronger strategic position to organize and change the outer world than is a society that generally sees things quite differently. A vantage point that understands humanity as small within the whole, as constantly confronting a great variety of forces,[14] and a perspective that concentrates on maintaining harmonious unity is arguably not as well situated to affect its surroundings in the ways that this "age of development's" policies and programs have done.[15] Climate change and resource wars resulting from hyper-development, however, now should prompt ponderings about which view may offer the most helpful approach for the future.

Fortunately, philosophically and theologically substantive efforts at qualitative rethinking are underway and their impacts on the narratives dominant heretofore are being felt. To that hopeful end, the African continent on the whole offers much of value for instruction. The story of Zambian miners' encounter with the forces of modern development theory in action represents only one bit of the rich wisdom and insight available there.

Notes

1. See the discussion of Pepper's worldview terms in chapter one.

2. Taylor and Lehmann offer a somewhat similar observation about the expansiveness and flexibility of faith experiences as observed in Copperbelt residents. "So faith ebbs and flows with the sense of psychic well-being; and if one way more than another can produce the assurance of being right with God, a man will turn to it." *Christians of the Copperbelt*, 290.

3. Taylor and Lehmann use the term "spiritual insecurity." *Christians of the Copperbelt*, 290.

4. Louis Dupré, "A Fateful Separation of Philosophy and Theology," *Harvard Divinity Bulletin* 35, nos. 2 and 3 (Spring/Summer 2007): 51-59.

5. Deneulin, *Religion in Development*, 52.

6. This was the case with one mining executive who wondered how a miner who called himself a Christian could leave an injured workmate to die underground. Another expatriate executive once mused, "What is really frustrating is that this is a Christian society . . . and what's frustrating about it are—not being honest and having the respect and all those [things] which are Christian virtues. An example: I could have a guy come in here, take my computer, throw it out the window and I'll tell him he's a very stupid idiot and that becomes more of an issue than the fact that he's destroyed company property. . . . Because here they've gone and destroyed something or even stolen it. . . . If I said,

'You—' and used colorful language, that [property destruction incident] would be forgotten and the colorful language would be the issue on the table." Interview with author, 2004.

7. The convening of the World Faiths Development Dialogue and the work of former Bank official Katherine Marshall serve as the most potent examples of institutional rethinking in this regard. See, for example, Tyndale, "The World Faiths Development Dialogue" and Marshall and Van Saanen, *Development and Faith.* Even so, the effort is monumental as the title of a couple of key scholarly articles on the subject illustrate. Wendy Tyndale, "Faith and Economics in 'Development': A Bridge Across the Chasm?" *Development in Practice* 10, no. 1 (February 2000): 9-18; Kurt Allen Ver Beek, "Spirituality: A Development Taboo," *Development in Practice* 10, no. 1 (February 2000): 31-43. A reflexive response by development policy makers from a secular stance might be that the proposals made in this Copperbelt study are naïve or unserious. Yet, are they any more naïve or unserious than structuring efforts billed as benevolent for societies in ways that actually do not accommodate how those societies make meaning?

8. Deneulin, *Religion in Development*, 4.

9. Some anthropological assessments of religion such as that offered by Talal Asad and discussed in Deneulin assert that symbols cannot be assumed to prompt specific motivations. "It is not because someone wears the symbol of the cross that s/he will be disposed to live out the Bible's teaching of loving one's enemy." *Religion in Development*, 58. The Copperbelt case potentially poses a compelling counter argument to this view.

10. Sen, *Development as Freedom*, 3ff.

11. Chinweizu, *Decolonizing the African Mind*, 13-30.

12. Some exceptions to this pattern might be found among Zambians who had been sent outside the country for training during earlier eras. But it is important to recall that the system of colonial era training was not designed with the long term intention of having Zambians take full control of the mines. Nor were the international training efforts of the nationalized era able to accommodate everyone who might have most benefited from the experience. See the discussions of social circumstances in chapter three as well as in Chinweizu, *Decolonizing the African Mind.*

13. See arguments in support of this proposition as made in Memmi, *The Colonizer and the Colonized* and Davidson, *The Black Man's Burden.*

14. "Any occurrence may be significant for [a person's] well-being and therefore needs to be accounted for in terms of some personal will, either of the living or the dead. Life becomes an unceasing Who-goes-there?, and a man alternates between triumph and anxiety." John V. Taylor, *The Primal Vision: Christian Presence Amid African Religion* (London: SCM Press, 1975), 71.

15. These observations resonate with what Deneulin terms an "instrumental vision of religion" in regard to development and as closely associated with Weberian analysis. But they are also intended to be compatible with the notion of religion as a moral force—a viewpoint that Deneulin argues deserves more attention by the professional development community. See the entire second chapter of *Religion in Development.* The point being made here is that both ways of understanding are necessary, particularly for helping well-intentioned Americans who wish to examine their own assumptions on these matters.

♀

Bibliography

Books, Chapters, and Journal Articles

Abram, David. *The Spell of the Sensuous: Perception and Language in a More-Than-Human World.* New York: Vintage Books, 1996.

Achebe, Chinua. *Things Fall Apart.* New York: Doubleday and Co., 1994.

Alam, M. Shahid. "Global Disparities Since 1800: Trends and Regional Patterns." *Journal of World Systems Research* XII, no. 2 (July 2006): 37-59.

Alverson, Hoyt. *Mind in the Heart of Darkness: Value and Self-Identity Among the Tswana of Southern Africa.* New Haven, CT: Yale University Press, 1978.

Apostel, Leo. *African Philosophy: Myth or Reality?* Gent, Belgium: Scientific Publishers, 1981.

Appiah, Kwame Anthony. *In My Father's House: Africa in the Philosophy of Culture.* Oxford: Oxford University Press, 1993.

Armstrong, Karen. *A History of God: The 4,000 Year Quest of Judaism, Christianity and Islam.* New York: Ballatine Books, 1993.

Arvidson, Malin. "Contradictions and Confusions in Development Work: Exploring the Realities of Bangladeshi NGOs." *Journal of South Asian Development* 3 (2008): 109-34.

Ashforth, Adam. "An Epidemic of Witchcraft? The Implications of AIDS for the Post-Apartheid State." *African Studies* 60, no. 1 (2002): 121-43.

———. *Madumo: A Man Bewitched.* Cape Town: David Philip Publishers, 2000.

———. *Witchcraft, Violence, and Democracy in South Africa.* Chicago: University of Chicago Press, 2005.

Aswathanarayana, U. *Mineral Resources Management and the Environment.* Exton, PA: A. A. Balema Publishers, 2003.

Austen, Ralph A. "The Moral Economy of Witchcraft: An Essay in Comparative History." Pp. 89-100 in *Modernity and Its Malcontents: Ritual and Power in Postcolonial Africa,* edited by Jean Comaroff and John Comaroff. Chicago: University of Chicago Press, 1993.

Balcomb, Anthony O. "Modernity and the African Experience." *Bulletin for Contextual Theology in Southern Africa* 3, no. 2 (1995): 12-20.

———. "Of Radical Refusers and Very Willing Victims—Interpolations of the Missionary Message in the Stories of Nongqawuse, Nxele, Ntsikana, and

Soga." *Bulletin for Contextual Theology in Southern Africa* 5, no. 1 (1997): 4-15.

———. "The Power of Narrative: Constituting Reality through Storytelling." Pp. 49-62 in *Orality, Memory and the Past: Listening to the Voices of Black Clergy Under Colonialism and Apartheid*, edited by Philippe Denis. Pietermaritzburg: Cluster Publications, 2000.

———. "Sex, Sorcery, and Stigma—Probing Some No-Go Areas of the Denial Syndrome in the AIDS Debate." *Journal of Theology for Southern Africa* 1, no. 125 (July 2006): 104-14.

———. "Rediscovering the Shock of the Other's Gaze—Constructing Identity in the Era of Postcolonial and Post-Apartheid Africa and in a Globalized World." *Theotherjournal.com: an intersection of theology and culture* 6, (2005): http://www.theotherjournal.com/article.php?id=87 (accessed March 29, 2009).

Bales, Kevin. *Disposable People: New Slavery in the Global Economy*. 4th ed. Berkeley: University of California Press, 2004.

Baran, Paul. *The Political Economy of Growth*. New York: Prometheus Paper Back, 1957.

Bates, Robert H. *Patterns of Uneven Development: Causes and Consequences in Zambia*. Monograph Series in World Affairs 11, no. 3. Denver: University of Denver, 1974.

Bediako, Kwame. *Jesus in Africa: The Christian Gospel in African History and Experience*. Akropong-Akupem, Ghana: Regnum Africa, 2000.

———. *Theology and Identity: The Impact of Culture Upon Christian Thought in the Second Century and in Modern Africa*. Oxford: Regnum Books, 1992.

Berger, Peter L., and Thomas Luckmann. *The Social Construction of Reality: A Treatise in the Sociology of Knowledge*. New York: Anchor Doubleday, 1966.

Binsbergen, Wim M. J. van. "Explorations in the History and Sociology of Territorial Cults in Zambia." Pp. 47-88 in *Guardians of the Land*, edited by J. M. Schoffeleers. Gwelo: Mambo Press, 1979.

———. *Religious Change in Zambia: Exploratory Studies*. London: Kegan Paul International, 1981.

Boahen, A. Adu. *African Perspectives on Colonialism*. Baltimore: Johns Hopkins University Press, 1987.

Bonk, Jonathan J. *Missions and Money: Affluence as a Missionary Problem . . . Revisited*. Maryknoll, NY: Orbis Books, 2006.

Bornstein, Erica. *The Spirit of Development: Protestant NGOs, Morality and Economics in Zimbabwe*. New York: Routledge, 2003.

Bosch, David J. *Transforming Mission: Paradigm Shifts in Theology of Mission*. Maryknoll, NY: Orbis Books, 1991.

Bostock, Mark, and Charles Harvey. *Economic Independence and Zambian Copper: A Case Study of Foreign Investment*. New York: Praeger Publishers, 1972.

Bradley, Kenneth. *Copper Venture: The Discovery and Development of Roan Antelope and Mufulira.* Great Britain: Mufulira Mines, Ltd., and Roan Antelope, Ltd., 1952.

Brockman, Norbert C. "Lenshina Mulenga Mubisha, Alice." *Dictionary of African Christian Biography*, 1994 http://www.dacb.org/stories/zambia/lenshina1_alice.html (accessed December 31, 2009).

Brooks, A. S. *Roan Antelope: A History of the Discovery and Development of the Roan Antelope Copper Mine.* N.p.: Luanshya Division, 1976.

Buell, Raymond Leslie. *The Native Problem in Africa.* Cambridge, MA: Bureau of International Research of Harvard University and Radcliffe College, 1928.

Bulpin, T. V. *Trail of the Copper King.* Cape Town: Howard B. Timmins, 1959.

Burawoy, Michael. *The Colour of Class on the Copper Mines: From African Advancement to Zambianization.* Zambian Papers 7. Lusaka: University of Zambia Institute for African Studies, 1972.

Campbell White, Oliver, and Anita Bhatia. *Privatization in Africa.* Washington, DC: The World Bank, 1998.

Césaire, Aimé. "Discourse on Colonialism." Pp. 222-27 in *African Philosophy: An Anthology*, edited by E. Chukwudi Eze. Malden, MA: Blackwell Publishers, 1998.

Chadwick, Owen. *MacKenzie's Grave.* London: Hodder and Stoughton, 1959.

Chambers, Robert. *Whose Reality Counts? Putting the Last First.* London: Intermediate Technology Publications, 1997.

Chidester, David. *Christianity: A Global History.* San Francisco: HarperSanFrancisco, 2000.

Chinweizu. *Decolonizing the African Mind.* Lagos: Pero Press, 1987.

Clarke, Gerard, and Michael Jennings, eds. *Development, Civil Society and Faith-Based Organizations: Bridging the Sacred and the Secular.* Basingstoke: Palgrave Macmillan, 2008.

Cline, Walter. "The Sources of Metals and Techniques of Metal Working in Negro Africa." PhD diss., Harvard University, 1936.

Comaroff, Jean, and John L. Comaroff, eds. *Modernity and its Malcontents: Ritual and Power in Postcolonial Africa.* Chicago: University of Chicago Press, 1993.

Cooper, Frederick. *Africa Since 1940: The Past of the Present.* Cambridge: Cambridge University Press, 2002.

Copetas, A. Craig. *Metal Men: Marc Rich and the 10-Billion-Dollar Scam.* New York: G. Putnam's Sons, 1985.

Craig, John. "Putting Privatisation into Practice: The Case of Zambia Consolidated Copper Mines Limited." *The Journal of Modern African Studies* 39, no. 3 (September, 2001): 389-410.

Daly, Herman E., and John B. Cobb, Jr., *For the Common Good: Redirecting the Economy Toward Community, the Environment and a Sustainable Future.* 2nd ed. Boston: Beacon Press, 1994.

Davidson, Basil. *The Black Man's Burden—Africa and the Curse of the Nation State.* New York: Times Books, 1992.

Davis, J. Merle. *Modern Industry and the African.* 2nd ed. London: Frank Cass and Co., Ltd., 1967.

de Gruchy, Steve. "Like Moses, We Always Perish Outside the Promised Land: Reinhold Niebuhr and the Contribution of Theology to Development." Pp. 133-50 in *Theology in Dialogue: The Impact of the Arts, Humanities, and Science on Contemporary Religious Thought*, edited by Lyn Holmes and Ralf K. Wustenberg. Grand Rapids, MI: Eerdmanns, 2002.

Deneulin, Séverine, with Masooda Bano. *Religion in Development: Rewriting the Secular Script.* London: Zed Books, 2009.

Dickson, Kwesi, and Paul Ellingworth. *Biblical Revelation and African Beliefs.* London: Lutterworth Press, 1969.

Dupré, Louis. "A Fateful Separation of Philosophy and Theology." *Harvard Divinity Bulletin* 35, nos. 2 and 3 (Spring/Summer 2007): 51-59.

Earle, Lucy. "International NGOs and Indigenous Social Movements." Policy Briefing Paper 15, January 2007 http://www.intrac.org/pages/policy_briefing_papers.html (accessed March 29, 2009).

Eisler, Riane. *The Real Wealth of Nations: Creating a Caring Economics.* San Francisco: Berrett-Koehler Publishers, Inc., 2007.

Ellis, Stephen, and Gerrie Ter Haar. *Worlds of Power: Religious Thought and Political Practice in Africa.* London: Hurst and Co., 2004.

Epstein, A. L. *Politics in an Urban African Community.* Manchester: Manchester University Press, 1958.

———. *Scenes from African Urban Life: Collected Copperbelt Papers.* Edinburgh: Edinburgh University Press, 1992.

Esteva, Gustavo. "Development." Pp. 6-25 in *The Development Dictionary: A Guide to Knowledge as Power*, edited by Wolfgang Sachs. London: Zed Books, 1992.

Etzioni, Amitai. *From Empire to Community.* New York: Palgrave MacMillan, 2004.

Evans-Pritchard, Edward E. *Witchcraft, Oracles and Magic Among the Azande.* Oxford: Clarendon Press, 1937.

Eze, E. Chukwudi, ed. *African Philosophy: An Anthology.* Malden, MA: Blackwell Publishers, 1998.

Fanon, Frantz. *Black Skins, White Masks.* Translated by Charles Lam Markmann. New York: Grove Press, Inc., 1967.

———. *The Wretched of the Earth.* 1st Evergreen Black Cat ed. New York: Grove/Atlantic, Inc., 1968.

Ferguson, James. *Expectations of Modernity: Myths and Meanings of Urban Life on the Zambian Copperbelt.* Berkeley: University of California Press, 1999.

Ferris, Elizabeth. "Faith-Based and Secular Human Organizations," *International Review of the Red Cross* 87, no.1 (June 2005): 311-25.

Fields, Karen E. *Revival and Rebellion in Colonial Central Africa.* Princeton: Princeton University Press, 1985.

Fixico, Donald L. "The Struggle for our Homes: Indian and White Values and Tribal Lands," Pp. 29-46 in *Defending Mother Earth: Native American Perspectives on Environmental Justice*, edited by Jace Weaver. Maryknoll, NY: Orbis Books, 1996.

Freire, Paulo. *Pedagogy of the Oppressed.* 30th anniversary ed. New York: Continuum, 2000.

Friedman, Thomas. *Hot, Flat, and Crowded: Why We Need a Green Revolution—and How it Can Renew America.* New York: Farrar, Straus, and Giroux, 2008.

Gardner, Gary T. *Inspiring Progress: Religions' Contributions to Sustainable Development.* New York: W. W. Norton and Co., 2006.

Geertz, Clifford. *The Interpretation of Cultures.* New York: Basic Books, 1973.

Gelfand, Michael. *Northern Rhodesia in the Days of the Charter: A Medical and Social Study 1878-1924.* Oxford: Basil Blackwell, 1961.

George, Susan. "How the Poor Develop the Rich." Pp. 207-213 in *The Post-Development Reader*, edited by Majid Rahnema and Victoria Bawtree. London: Zed Books, 1997.

Geschière, Peter. *The Modernity of Witchcraft: Politics and the Occult in Postcolonial Africa.* Charlottesville, VA: University Press of Virginia, 1997.

Gibbons, A. H. *Africa from South to North Through Marotseland.* Vol. 1. London: John Lane, 1904.

Gilkey, Langdon. *Nature, Reality, and the Sacred: The Nexus of Science and Religion.* Theology and the Sciences Series. Minneapolis: Fortress Press, 1993.

Gluckman, Max. *Custom and Conflict in Africa.* Oxford: Basil Blackwell, 1963.

Green, Duncan. *From Poverty to Power: How Active Citizens and Effective States Can Change the World.* Oxford: Oxfam, 2008.

Green, Edward C. "Purity, Pollution and the Invisible Snake in Southern Africa." *Medical Anthropology* 17 (1997): 83-100.

Greider, William. *The Soul of Capitalism: Opening Paths to a Moral Economy.* New York: Simon and Schuster, 2003.

Hackett Fischer, David. *Albion's Seed: Four British Folkways in America.* New York: Oxford University Press, 1989.

Hall, Richard. *Zambia.* London: Pall Mall Press, 1965.

———. *Zambia 1890-1964: The Colonial Period.* London: Longman, 1976.

Hallen, Barry, and J. O. Sodipo. *Knowledge, Belief and Witchcraft: Analytic Experiments in African Philosophy.* London: Ethnographica, 1986.

Hampden-Turner, Charles, and Alfons Trompenaars. *The Seven Cultures of Capitalism: Values Systems for Creating Wealth in the United States, Japan, Germany, France, Britain, Sweden, and the Netherlands.* New York: Currency Doubleday, 1993.

Hanna, A. J. *The Story of the Rhodesias and Nyasaland.* London: Faber and Faber, 1960.

Harper, Sharon, ed., *The Lab, the Temple, and the Market: Reflections at the Intersection of Science, Religion, and Development.* Bloomfield, CT: Kumarian Press, 2000.

Hayes, Steven C., Linda J. Hayes, and Hayne W. Reese. "Finding the Philosophical Core: A Review of Stephen C. Pepper's *World Hypotheses: A Study in Evidence.*" *Journal of the Experimental Analysis of Behavior* 50, no. 1 (July 1988): 97-111.

Haynes, Jeffrey. *Religion and Development: Conflict or Cooperation?* New York: Palgrave Macmillan, 2007.

Heisler, Helmuth. *Urbanisation and the Government of Migration: The Interrelation of Urban and Rural Life in Zambia.* London: C. Hurst and Co., 1974.

Henkel, Reinhard. *Christian Missions in Africa: A Sociological Study of the Impact of Their Activities in Zambia.* Berlin: Dietrich Reimer Verlag, 1989.

Herbert, Eugenia. *Red Gold of Africa: Copper in Precolonial History and Culture.* Madison, WI: University of Wisconsin Press, 1984.

Heron, Alastair. *Accidents in the Zambian Mining Industry.* Lusaka: Government Printer, 1968.

Hiebert, Paul. *Transforming Worldviews: An Anthropological Understanding of How People Change.* Grand Rapids, MI: Baker Academic, 2008.

Hilhorst, Dorothea. *The Real World of NGOs.* London: Zed Books, 2003.

Hinfelaar, Hugo F. "Witch-Hunting in Zambia and International Illegal Trade." Pp. 229-45 in *Imagining Evil: Witchcraft Beliefs and Accusations in Contemporary Africa*, edited by Gerrie Ter Haar. Trenton, NJ: Africa World Press, Inc., 2007.

Holland, Heidi. *African Magic: Traditional Ideas that Heal a Continent.* London: Viking, 2001.

Hollaway, John. *All Poor Together: The African Tragedy and Beyond.* Johannesburg: Capricorn Books, 2000.

Horton, Robin. *Patterns of Thought in Africa and the West: Selected Theoretical Papers in Magic, Religion, and Science.* Cambridge: Cambridge University Press, 1993.

Hovland, Ingie. "Who's Afraid of Religion? Tension Between 'Mission' and 'Development' in the Norwegian Mission Society." Pp. 171-86 in *Development, Civil Society and Faith-Based Organizations: Bridging the Sacred and the Secular*, edited by Gerard Clarke and Michael Jennings. Basingstoke: Palgrave Macmillan, 2008.

Huntington, Samuel P. *The Clash of Civilizations and the Remaking of World Order.* London: Touchstone Books, 1996.

Jacobs, Donald R. "Culture and the Phenomena of Conversion," *Gospel in Context* 1, no. 3 (July 1978): 4-13.

Jahn, Janheinz. *Muntu: The New African Culture.* Translated by Marjorie Greene. New York: Grove Press, Inc., 1961.

James, William. *The Varieties of Religious Experience: A Study in Human Nature.* The Gifford Lectures 1901-1902. 1st Touchstone ed. New York: Touchstone Simon and Schuster, 1997.

Jamieson, Gladys. *Zambia Contrasts: A Page in the Story of Christian Education in Zambia.* London: London Missionary Society, 1965.

Kalu, Ogbu U. *Power, Poverty, and Prayer: The Challenges of Poverty and Pluralism in African Christianity, 1960-1996*. Trenton, NJ: Africa World Press, 2006.

Kaoma, Kapya John. "The Democratic Crisis in Southern Africa: A Challenge to the Church." unpublished book manuscript, Mutare, Zimbabwe, 2001.

Kaplan, Irving, et al. *Area Handbook for Zambia*. 2nd ed. Washington, DC: U.S. Government Printing Office, 1974.

———. ed. *Zambia: A Country Study*. Washington, DC.: The American University, 1979.

Kaunda, Francis. *Selling the Family Silver: The Zambian Copper Mines Story*. KwaZulu Natal: Interpak Books, 2002.

Kaunda, Kenneth D. *A Humanist in Africa: Letters to Colin M. Morris*. London: Longmans, Green and Co., Ltd., 1966.

Kay, George. *A Social Geography of Zambia: A Survey of Population Patterns in a Developing Country*. London: University of London Press, 1967.

Kelly, Marjorie. *The Divine Right of Capital: Dethroning the Corporate Aristocracy*. San Francisco: Berrett-Koehler Publishers, Inc., 2001.

Kirk, Andrew J. *What is Mission? Theological Explorations*. London: Darton, Longman, and Todd, 1999.

Kirwen, Michael E. *The Missionary and the Diviner: Contending Theologies of Christian and African Religions*. Maryknoll, NY: Orbis Books, 1987.

Klein, Naomi. *The Shock Doctrine: The Rise of Disaster Capitalism*. New York: Metropolitan Books, 2007.

Knitter, Paul, and Chandra Muzaffar, eds. *Subverting Greed: Religious Perspectives on the Global Economy*. Maryknoll, NY: Orbis Books, 2002.

Korten, David C. *Getting to the 21st Century: Voluntary Action and the Global Agenda*. West Hartford, CT: Kumarian Press, 1990.

———. *The Great Turning: From Empire to Earth Community*. West Hartford, CT: Kumarian Press, 2006.

———. *The Post-Corporate World: Life After Capitalism*. West Hartford, CT: Kumarian Press, 1999.

———. *When Corporations Rule the World*. West Hartford, CT: Kumarian Press, 1995.

Kuhn, Thomas S. *The Structure of Scientific Revolutions*. Chicago: University of Chicago Press, 1962.

Lagus, Charles. *Operation Noah*. London: William Kimber and Co., Ltd., 1959.

Landes, David. *The Wealth and Poverty of Nations*. London: Abacus, 1998.

Lane, James Eric. *Moment of Encounter*. New York: Peter Lang, 1984.

Lane, Stewart. *God Loves Africans to be African*. Limbe, Malawi: Cornelius Fellowship, 2000.

———. *Weep You Rich!* Limbe, Malawi: Cornelius Fellowship, n.d.

Larmar, Miles. *Mineworkers in Zambia: Labour and Political Change in Post-Colonial Africa*. London: I. B. Tauris and Co., 2006.

Lecky, William E. H. *The History of the Rise and Influence of the Spirit of Rationalism in Europe. Vol. 1*. London: Watts and Co., 1910.

Levi-Strauss, Claude. *The Savage Mind.* Chicago: University of Chicago Press, 1962.
Linden, Ian. "Globalization and the Church: An Overview." Pp. 3-11 in *Development Matters: Christian Perspectives on Globalization*, edited by Charles Reed. London: Church House Publishing, 2001.
MacDonald, Paul S. *Descartes and Husserl: The Philosophical Project of Radical Beginnings*. Albany: State University of New York Press, 2000.
Maluleke, Tinyiko Sam. "A Rediscovery of the Agency of Africans." Pp. 165-90 in *Marginalized Africa: An International Perspective*, edited by Peter Kanyandago. Nairobi: Paulines Publications, 2002.
Manchishi, P. C. "The Growth of Teacher Education in Zambia Since Independence." *The African Symposium* 4, no. 4 (December 2004): 2-22.
Marshall, Katherine. "Development and Religion: A Different Lens on Development Debates." *Peabody Journal of Education* 76, no. 3 (2001): 339-75.
Marshall, Katherine, and Marisa Bronwyn Van Saanen, eds. *Development and Faith: Where Mind, Heart, and Soul Work Together*. Washington, DC: The World Bank, 2007.
Martin, Antony. *Minding Their Own Business: Zambia's Struggle Against Western Control.* London: Hutchinson and Co., 1972.
Matongo, Albert B. K. "Popular Culture in a Colonial Society: Another Look at Mbeni and Kalela Dances on the Copperbelt, 1930-64." Pp. 180-217 in *Guardians in Their Time: Experiences of Zambians Under Colonial Rule, 1890-1964*, edited by Samuel N. Chipungu. London: Macmillan Press, 1992.
Mazrui, Ali. *The African Connection.* London: Cambridge University Press, 1980.
Mbiti, John. *African Religions and Philosophy*. London: Heinemann, 1969.
McDaniel, Charles. *God and Money: The Moral Challenge of Capitalism*. Lanham, MD: Rowman and Littlefield Publishers, Inc., 2007.
McFague, Sallie. "God's Household: Christianity, Economics, and Planetary Living." Pp. 119-36 in *Subverting Greed: Religious Perspectives on the Global Economy*, edited by Paul Knitter and Chandra Muzzafar. Maryknoll, NY: Orbis Books, 2002.
Meebelo, Henry S. *Reaction to Colonialism. A Prelude to the Politics of Independence in Northern Zambia 1893-1939*. Manchester: Manchester University Press, 1971.
Memmi, Albert. *The Colonizer and the Colonized.* Translated by Howard Greenfield. First American ed. New York: Orion Press, 1965. Boston: Beacon Press, 1991. Page references are to the 1991 edition.
Mendelsohn, Felix, ed. *The Geology of the Northern Rhodesian Copperbelt.* London: MacDonald, 1961.
Mihevc, John. *The Market Tells Them So: The World Bank and Economic Fundamentalism in Africa.* London: Zed Books, 1995.
Mitchell, J. Clyde. *The Kalela Dance—Aspects of Social Relationships Among Urban Africans in Northern Rhodesia.* Rhodes-Livingstone Paper 22. Livingstone: Rhodes-Livingstone Institute, 1956.

Moore, Henrietta L., and Todd Sanders. *Magical Interpretations, Material Realities: Modernity, Witchcraft, and the Occult in Postcolonial Africa.* London: Routledge, 2001.

Moyo, Dambisa. *Dead Aid: Why Aid is Not Working and How There is Another Way for Africa.* London: Allen Lane, 2009.

M'Passou, Denis. *Mindolo: A Story of the Ecumenical Movement in Africa.* Lusaka: Multimedia Press, 1983.

Mtetwa, Sipho. "African Spirituality in the Context of Modernity," *Bulletin for Contextual Theology in Southern Africa* 3, no. 2 (1996): http://sorat.ukzn.ac.za/theology/bct/tonyb.htm (accessed April 22, 2009).

Mudimbe, V. Y. *The Invention of Africa: Gnosis, Philosophy, and the Order of Knowledge.* Bloomington, IN: Indiana University Press, 1988.

Mueller, Dennis C., ed. *The Political Economy of Growth.* New Haven: Yale University Press, 1983.

Mutwa, Credo. *Indaba, My Children: African Tribal History, Legends, Customs and Religious Beliefs.* Edinburgh: Payback Press, 1998.

Myers, Bryant. *Walking with the Poor: Principles and Practices of Transformational Development.* Maryknoll, NY: Orbis Books, 1999.

Neusner, Jacob. *The Way of Torah: An Introduction to Judaism.* Belmont, CA: Wadsworth Publishing Co., 1993.

Niebuhr, H. Richard. *Christ and Culture.* San Francisco: Harper Collins, 1951.

———. *Radical Monotheism and Western Culture.* New York: Harper and Row, 1960.

Neill, Stephen. *A History of Christian Missions.* 2nd ed. New York: Viking Penguin, 1986.

Nisbett, Richard E. *The Geography of Thought: How Asians and Westerners Think Differently . . . and Why.* New York: The Free Press, 2003.

Nürnberger, Klaus. *Beyond Marx and Market: Outcomes of a Century of Economic Experimentation.* Pietermaritzburg: Cluster Publications, 1998.

Occhipinti, Laurie A. *Acting on Faith: Religious Development Organizations in Northwestern Argentina.* Lanham, MD: Lexington Books, 2005.

Oliver, Roland, and J. D. Fage. *A Short History of Africa.* London: Penguin Books, 1995.

Pallister, David, Sarah Stewart, and Ian Lepper. *South Africa Inc.: The Oppenheimer Empire.* London: Simon and Schuster, 1987.

Parpart, Jane. "Class and Gender on the Copperbelt: Women in Northern Rhodesian Copper Mining Areas 1926-1964." Working Paper No. 77, African Studies Center, Boston University, 1983.

Parsons, Elizabeth C. "Insights from Zambian Miners for Rethinking Development Policy and the Environment." *Development in Practice* 18, no. 3 (June 2008): 403-11.

———. "Provoking the Rocks: A Study of Reality and Meaning on the Zambian Copperbelt." PhD diss., University of KwaZulu-Natal, 2007.

Pepper, Stephen C. *World Hypotheses: A Study in Evidence.* Berkeley: University of California Press, 1942.

Powdermaker, Hortense. *Copper Town: Changing Africa; The Human Situation on the Rhodesian Copperbelt.* New York: Harper and Row, 1962.

Quarles van Ufford, Philip, and Matthew Schoffeleers. *Religion and Development: Towards an Integrated Approach.* Amsterdam: Free University Press, 1988.

Rahnema, Majid, and Victoria Bawtree, eds. *The Post-Development Reader.* London: Zed Books, 1997.

Rasmussen, Larry. *Earth Community Earth Ethics.* Maryknoll, NY: Orbis Books, 1996.

Ripley, Earle A., Robert E. Redmann, and Adèle A. Crowder. *Environmental Effects of Mining.* Delray Beach, FL: St. Lucie Press, 1996.

Rist, Gilbert. *The History of Development from Western Origins to Global Faith.* Translated by Patrick Camiller. London: Zed Books, 1997.

Robert, Dana. *Christian Mission: How Christianity Became a World Religion.* Malden, MA: Wiley-Blackwell, 2009.

Roberts, Andrew. *A History of Zambia.* New York: Africana Publishing Co., 1976.

Rotberg, Robert I. *Christian Missionaries and the Creation of Northern Rhodesia 1880-1924.* Princeton: Princeton University Press, 1965.

Sachs, Wolfgang, ed. *The Development Dictionary: A Guide to Knowledge as Power.* London: Zed Books, 1992.

Sarah, Archbishop Robert. *Culture, Democracy, and Development in the Light of Centesemus Annus.* Nairobi: Paulines Publications, 2000.

Schutz, Alfred. *Collected Papers Vol. I, The Problem of Social Reality.* Edited by Maurice Natanson. The Hague: Martinus Nijhoff, 1962.

———. *Collected Papers Vol. II, Studies in Social Theory.* Edited by Arvid Brodersen. The Hague: Martinus Nijhoff, 1976.

———. *Collected Papers Vol. III, Studies in Phenomenological Philosophy.* Edited by Ilse Schutz. The Hague: Martinus Nijhoff, 1975.

———. *Collected Papers Vol. IV.* Edited by Helmut Wagner and George Psathas in collaboration with Fred Kersten. Dordrecht: Kluwer Academic Publishers, 1996.

———. "On Multiple Realities." Pp. 207-59 in *Collected Papers Vol. 1, The Problem of Social Reality*, edited by Maurice Natanson. The Hague: Martinus Nijoff, 1962.

———. "The Stranger." Pp. 91-105 in *Collected Papers Vol. 2, Studies in Social Theory*, edited by Arvid Brodersen. The Hague: Martinus Nijhoff, 1976.

Schutz, Alfred, and Thomas Luckmann. *The Structures of the Life-World.* Translated by Richard M. Zaner and H. Tristram Engelhardt, Jr. Evanston, IL: Northwestern University Press, 1973.

Schwab, Peter. *Africa: A Continent Self-Destructs.* New York: Palgrave Macmillan, 2001.

Scott, James. *Weapons of the Weak: Everyday Forms of Peasant Resistance.* New Haven: Yale University Press, 1985.

Sen, Amartya. *Development as Freedom.* New York: Alfred A. Knopf, 1999.

Serequeberhan, Tsenay. "Colonialism and the Colonized: Violence and Counter-Violence." Pp. 234-54 in *African Philosophy: An Anthology*, edited by E. Chukwudi Eze. Malden, MA: Blackwell Publishers, 1998.

Silavwe, Geoffrey W. *Some Aspects of Personnel Management Practice in the Copper Mining Industry of Zambia*. Ndola: Mission Press, 1999.

Simmel, Georg. *The Sociology of Georg Simmel*. Translated and edited by Kurt H. Wolff. Glencoe, IL: The Free Press, 1950.

Stiff, Peter. *See You in November*. Alberton: Galago Publishing, Inc., 1985.

Stigand, Chauncey Hugh. *Hunting the Elephant in Africa and Other Wanderings*. 1913. African Hunting Reprint Series Vol. 8. Bulawayo: Books of Zimbabwe, 1989.

Stone, Merlin. *When God Was a Woman*. New York: Harcourt Brace Jovanovich, 1973.

Taylor, John V. *The Primal Vision: Christian Presence Amid African Religion*. London: SCM Press, 1975.

Taylor, John V., and Dorothea Lehmann. *Christians of the Copperbelt: The Growth of the Church in Northern Rhodesia*. London: SCM Press, 1961.

Tempels, Placide. *Bantu Philosophy*. Translated by Colin King. Paris: Présense Africaine, 1959.

Tinker, George. "An American Indian Theological Response to Ecojustice." Pp. 153-76 in *Defending Mother Earth: Native American Perspectives on Environmental Justice*, edited by Jace Weaver. Maryknoll, NY: Orbis Books, 1996.

Tocqueville, Alexis de. *Democracy in America*. 1835 and 1840. Edited and abridged by Richard Heffner. New York: Signet Classic, 2001.

Tucker, Mary Evelyn, and John A Grim, eds. *Worldviews and Ecology: Religion, Philosophy, and the Environment*. Maryknoll, NY: Orbis Books, 1994.

Tyndale, Wendy. "Faith and Economics in 'Development': A Bridge Across the Chasm?" *Development in Practice* 10, no. 1 (February 2000): 9-18.

———. "The World Faiths Development Dialogue: An Interfaith Response to Poverty and Development in an Age of Globalization." Pp. 106-15 in *Development Matters: Christian Perspectives on Globalization*, edited by Charles Reed. London: Church House Publishing, 2001.

Veltmeyer, Henry. *Illusion or Opportunity: Civil Society and the Quest for Social Change*. Halifax: Fernwood Publishing, 2007.

Ver Beek, Kurt Alan. "Spirituality: A Development Taboo." *Development in Practice* 10, no. 1 (February 2000): 31-43.

Wariboko, Nimi. *The Depth and Destiny of Work: An African Theological Interpretation*. Trenton: Africa World Press, 2008.

———. *God and Money: A Theology of Money in a Globalizing World*. Lanham, MD: Lexington Books, 2008.

Watson, William. *Tribal Cohesion in a Money Economy: A Study of the Mambwe People of Northern Rhodesia*. Manchester: Manchester University Press, 1958.

Weber, Max. *The Protestant Ethic and the Spirit of Capitalism*. 1904. Translated by Talcott Parsons. 2nd Roxbury ed. Los Angeles: Roxbury Publishing Co., 1998.

Whaites, Alan. "NGOs, Civil Society, and the State: Avoiding Theoretical Extremes in Real World Issues." *Development in Practice* 8, no. 3 (1998): 343-49.

White, Luise. *Speaking with Vampires: Rumor and History in Colonial Africa*. Berkeley: University of California Press. 2000.

White, Sarah, and Romy Tiongco. *Doing Theology and Development: Meeting the Challenge of Poverty*. Edinburgh: Saint Andrews Press, 1997.

Wilson, Godfrey, and Monica Hunter. *The Study of African Society*. Rhodes-Livingstone Paper 2. Livingstone: Rhodes-Livingstone Institute, 1939.

Wiredu, Kwasi. "How Not to Compare African Thought with Western Thought." Pp. 193-99 in *African Philosophy: An Anthology*, edited by E. Chukwudi Eze. Malden, MA: Blackwell Publishers, 1998.

Wood, Anthony St. John. *Northern Rhodesia: The Human Background*. London: Pall Mall Press, 1961.

Wright, Ronald. *What is America? A Short History of the New World Order*. Toronto: Vintage Canada, 2008.

Wuthnow, Robert. *Boundless Faith: The Global Outreach of American Churches*. Berkeley: University of California Press, 2009.

Zambia Episcopal Conference, Christian Council of Zambia, and Zambia Evangelical Fellowship. *Marxism, Humanism, and Christianity*. Lusaka: Teresianum Press, 1979.

Archival Material, Newspaper Articles, Presentations, Online Sources, and Reports

"AF Plane Crash Report Inquiries Concluded." *Times of Zambia*, July 19, 2005, http://allafrica.com/stories/printable/200507190340.html (accessed July 20, 2005).

Annual Report of the Mines Department for the Year 1968. Lusaka: Government Printer, 1969.

Annual Report of the Mines Department for the Year 1969. Lusaka: Government Printer, 1970.

Balcomb, Anthony O. "Descartes Meets the *Isangoma*—The Encounter Between Modern and Indigenous Knowledge Systems Beyond Colonialism and Apartheid." Presentation delivered at Indigenous Knowledge Conference, Saskatoon, Saskatchewan, April 2001.

———. "Science and the African Worldview—Rediscovering the Numinous, Re-Animating the Cosmos." Presentation delivered at University of Cape Town, Cape Town, April 2003.

"Binani Wins Bid." *Mining Mirror*, July 1997, 7.

Boston University School of Education. International Educational Development Program. http://www.bu.edu/sed/programsinternational.htm (accessed December 29, 2009).

Center for International Development at Harvard University http://www.cid.harvard.edu/cidinformation/ (accessed December 29, 2009).

Chen, Yulai. "China in Africa." Presentation delivered at African Studies Center, Boston University, Boston, MA, November 23, 2009.

"Chilubi Couple Suspected as 'Human Crocodiles' Killed." *Zambia Daily Mail*, February 7, 2006, 2.

"Christians Pray for Peaceful Zambia." *Sunday Times*, May 23, 2005.

Commission of Inquiry Appointed by His Excellency, the President of the Republic of Zambia, Dr. Kenneth David Kaunda. *The Mufulira Mine Disaster: Final Report on the Causes and Circumstances of the Disaster Which Occurred at Mufulira Mine on the 25 September 1970.* Lusaka, 1971.

"Copper and the Copper Mining Industry of This Country." Photocopy, ZCCM Archives, Ndola, October 1, 1964.

"Copper Studies." *CRU Analysis* 32, no. 11 (May, 2005): 4.

"Copperbelt Energy's Changing of Guard." *Sunday Times of Zambia*, March 5, 2006, 8 "Danger That Went Unseen: Muf Inquiry Reveals Wrong Diagnosis." *Zambia Daily Mail*, February 18, 1971.

"Do Not Gain from Disasters." *Zambia Daily Mail*, May 27, 2005.

Dymond, Abi, Kato Lambrechts, and Simon Chase, "Undermining Development? Copper Mining in Zambia." Report sponsored by Action for Southern Africa, Christian Aid, and SCAIF 2007. http://www.actsa.org/Pictures/UpImages/pdf/Undermining%20development%20report.pdf 2007 (accessed March 23, 2009).

"Enhancing Economic Diversification in Zambia's Copperbelt." Report of a Fact Finding Mission to the Copperbelt by Private Sector Donor Group August 14-16, 2002. Lusaka, 2003.

"Environmental Awareness Centre Construction Completed." *Sunday Times of Zambia*, December 5, 2004, 1.

"Enya Holdings to Operate Zambia's Top Cobalt Producer." Miningreview.com, June 26, 2009. http://beta.miningreview.com/node/15730 (accessed December 5, 2009).

"Episcopalians for Global Reconciliation." http://www.e4gr.org/ (accessed March 25, 2009).

"Evil of Casualisation, The." *The Post*, May 22, 2005, 23.

Fraser, Alastair and John Lungu. "For Whom the Wind Falls: Winners and Losers in the Privatisation of Zambia's Mines." Report sponsored by Civil Society Trade Network of Zambia and Catholic Centre for Justice, Development and Peace, Zambia Episcopal Conference, 2006. http://www.minewatchzambia.com/reports/report.pdf (accessed March 23, 2009).

Gulhati, Ravi. "Impasse in Zambia: The Economics and Politics of Reform." EDI Development Policy Case Series, Analytical Case Studies no. 2. Washington, DC.: The World Bank, 1989.

Hobson, Dick "ZCCM, the Story That Began at Matero About 13 Years Ago." *Mining Mirror*, March 26, 1982, 10.

"How the Mines Sustain Our Economy." *Mining Mirror*, October 22, 1984, 3.

Intergovernmental Panel on Climate Change. "Climate Change 2007: Synthesis Report." Presented at IPCC Plenary XXVII, Valencia, Spain, November 2007. http://www.ipcc.ch/pdf/assessmentreport/ar4/syr/ar4_syr_spm.pdf (accessed October 6, 2008).

"International Development Association Program Document for a Proposed Credit in the Amount of SDR 27.4 Million (US$40 Million Equivalent) to the Republic of Zambia for an Economic and Growth Management Credit." Report 29294 ZA. Washington, DC: The World Bank, 2004.

"Introduction to the Mining Industry for Students." Ndola: Roan Selection Trust, 1979.

Jubilee Research. http://www.jubileeresearch.org/ (accessed April 20, 2009).

"Kafue Deal Collapse May Delay Privatization." *Sunday Times of Zambia*, April 5, 1998. http://www.btimes.co.za/98/0405/comp/comp4.htm (accessed January 7, 2010).

"Kafumukache: Refined Politician." *Zambia Daily Mail*, December 9, 2004, 7.

Kaleyi, Calvin. "Copper Crisis in Zambia." TradersAfrica.com, August-November, 2002. http://www.tradersafrica.com/articles.asp?articleid={EB470C44-E005411198DE-69D21DE0ER3F} (accessed January 7, 2010).

Kaunda, Kenneth D. "Humanism in Zambia and a Guide to its Implementation Part 1." Lusaka: Government Printer, 1968.

———. "The Imperative of Human Dignity." Address to the Inaugural Session of the Non-Governmental Organisations International Conference on Human Rights, Paris, September 15, 1968. (Lusaka: Zambia Information Services, 1968).

Lembalemba, Kaunda. Interview on "Special Report." ZNBC-TV, May 20, 2005.

"Lembalemba Puts Blame on Mopani Management for Accidents at Nkana." *The Post*, June 27, 2005. http://allafrica.com/stories/printable/200506271215.html (accessed July 3, 2005).

"Let's Grow More Food, Appeals Shapi." *Mining Mirror*, October 22, 1984, 1.

"Let's pray at all times—Levy." *Zambia Daily Mail*, May 24, 2005, 1.

"Levy Urges Zambians to Approach God at All Times." *The Post*, May 23, 2005.

"Luanshya, Chibuluma Sale Pact Sealed." *Mining Mirror*, September 1997, 1.

McNeil, Jr., Donald. "For Sale: Zambia's Rich Copper Mines, All of Them." *New York Times*, August 11, 1996. http://www.nytimes.com/1996/08/11/world/for-sale-zambia-s-rich-copper-mines-all-of-them.html?pagewanted=1 (accessed January 7, 2010).

Millennium Development Goals. http://www.un.org/millenniumgoals/ (accessed October 6, 2008).

Mindolo Ecumenical Foundation. "Mindolo Work Camp." Minute book, World Council of Churches, Kitwe, July 10 - August 9, 1959.

Minejob.com. http://www.minejob.com/keywords-Metallurgy.html (accessed December 5, 2009).

"Mine Owners Should be Charged for Murder." *Zambia Daily Mail*, May 16, 2005, 2.

"Miners Want Muf Report Translated." *Zambia Daily Mail*, February 17, 1971.

"Mining Industry Under New Era." Photocopy, Chamber of Mines of Zambia, Kalulushi, 2004.

Ministry of Labour. "The Progress of Zambianization in the Mining Industry." Lusaka: Government Printer, 1968.

"Mopani Demands Government's Position Over its Fate." *The Post*, July 1, 2005. http://allafrica.com/stories/printable/200507010292.html (accessed July 3, 2005).

"Mufulira Mine Floods Again." *Times of Zambia*, March 29, 2005, 1.

Mukula, P. M. compiler. *Calendars of the District Notebooks (Copperbelt Province) 1899-1964. Vol. 4.* Lusaka: Government Printer, 1981.

National Institute for Personnel Research in Association with the Institute for Social Research. *The Attitudes of White Mining Employees Towards Life and Work on the Copperbelt.* Pietermaritzburg: University of Natal, November 1960.

"New Day, New Way: U.S. Foreign Assistance for the 21st Century." A Proposal from the Modernizing Foreign Assistance Network. 2008. http://www.cgdev.org/content/publications/detail/16210 (accessed October 6, 2008).

New Economics Foundation. http://www.neweconomics.org (accessed April 20, 2009).

"Notes on the Privatization of ZCCM." Press release, ZCCM Archives, Ndola, February 1992.

"No Zambian Should be Employed as Casual Worker—MUZ." *Times of Zambia*, May 12, 2005, 2.

"ONE campaign." http://www.one.org/ (accessed March 25, 2009).

"Only God Can Comfort Us—KK." *Zambia Daily Mail*, May 23, 2005, 1.

"Penza Weathers MUZ Storm." *Sunday Times*, September 23, 1992.

"Privatization on Course." *Mining Mirror*, December, 1997, 1.

"Privatization: Opportunities Still Abound." *The Investor Magazine* 2, (Quarter 1, 2002): 19-22.

"Public Inquest for BGRIMM Victims Fails to Take Off." *The Post*, July 7, 2005. http://allafrica.com/stories/printable/200507070875.html (accessed July 20, 2005).

Relly, Gavin W. H. "Background to the Zambian Copper Mining Industry." An Address by the Chairman of Anglo-American Corporation (Central Africa), Ltd. on the Producers Day at the Inter-governmental Copper Conference, Lusaka, 1967.

"Report of the Commission Appointed to Enquire into the Disturbances in the Copper belt, Northern Rhodesia Together with the Governor's Despatch to the Secretary of State on the Report, Herbert Young, Governer." N.p.: November 1935.

"Retired Miner Looks Back on 35 Year Career." *Mining Mirror*, February, 1995, 11.

"Ritual Deaths Rock Chiawa." *Sunday Times of Zambia*, December 5, 2004, 1.

"Roan Consolidated Mines, Ltd." Photocopy, ZCCM Archives, Ndola, 1980.

"Sikota Wina on Tuesdays: BGRIMM Explosives." *The Post*, April 26, 2005.

Sinyangwe, Chiwoyu and Ernest Chanda "Appreciate Workers' Efforts, Nchanga MP Urges KCM." *The Post*, November 16, 2009. http://www.postzambia.com/post-read_article.php?articleId=1945. (accessed December 5, 2009)

Situmbeko, Lishala G. and Jack Jones Zulu. "Zambia: Condemned to Debt. How the World Bank and IMF Have Undermined Development." London: World Development Movement, 2004.

"State Receives Report on Mine Accidents." *Zambia Daily Mail*, December 6, 2005.

Stiglitz, Joseph. "Globalism's Discontents." *The American Prospect*, Winter, 2002, A16-A21.

"Survey of the Zambian Mining Industry 1995-2004." Chamber of Mines of Zambia, Kalulushi, October 2005.

"Suspected Witch Crash-lands in Ndola." *Times of Zambia*, March 7, 2006, 2.

"Suspected Witches Find Refuge in Kaleni Hills." *Times of Zambia*, July 7, 2005, 9.

"Sylvester Nkhoma Recalls Good Old Days in the Mine." *Mining Mirror*, September, 1997, 7.

"Tough Days Ahead—KK." *Mining Mirror*, July 27, 1984, 1.

Towards Complete Independence: What it all Means. Lusaka: Government Printer, 1969.

"Traditional Healers Warn Witch-Finders Against Theft." *Daily Mail*, December 9, 2004, 3.

"Tribute to a Hard Working Miner." *Mining Mirror*, April, 1996, 5.

"Two Die in Mine Accident." *Zambia Daily Mail*, May 19, 2005. http://www.daily-mail.co.zm/front03.htm (accessed July 3, 2005).

"Turning an Economy Around: The Challenge in Zambia." OED Précis no. 132. Washington, DC: The World Bank, 1996.

"$25 Billion Question." *The Economist*, July 2, 2005, 24-26.

United for Change. http://www.muslimsunitedforchange.com/ (accessed March 25, 2009).

United Nations Millennium Development Goals (MDGs). http://www.un.org/millenniumgoals/ (accessed October 6, 2008).

United States Agency for International Development. "Prosperity, Hope and Better Health for Zambians." Country Strategic Plan FY2004-2010, Vol. 1: Strategy. June 2003. http://www.usaid.gov/zm/docs/csp_vol1.pdf (accessed March 29, 2009).

"Voices of the Poor." http://go.worldbank.org/IM4HTOJ690 (accessed on December 5, 2009).

Watts, W. G., ed. "Commemorative Programme, 7th Commonwealth Mining and Metallurgical Congress, Northern Rhodesia Section." Northern Rhodesia Section of the 7th Commonwealth Mining and Metallurgical Congress, Kitwe, 1961.

"Will Privatization Auction Zambia?" *Sunday Times of Zambia*, September 27, 1992.

"Witchcraft Act." *Zambia Criminal Law.* Vol 7, Ch. 90, 507-12. http://www.parliament.gov.zm/downloads/VOLUME%207.pdf (accessed March 25, 2009).

World Christian Database. http://www.worldchristiandatabase.org/ezproxy.bu.edu/wcd/ (accessed April 1, 2009).

World Faiths Development Dialogue. http://berkleycenter.georgetown.edu/wfdd (accessed on February 13, 2010).

"ZAAM Sale is Set in Motion." *Mining Mirror*, June, 1996, 1.

"Zam Threatens to Close Mines." *The Namibian*, June 27, 2005. www.namibian.com.na/2005/June/marketplace/05BE825C8F.html (accessed June 27, 2005).

Zambia Consolidated Copper Mines Limited. "Proposed Disposal of the Assets Forming the Mufulira Division and the Nkana Mines, Concentrator and Cobalt Plant and Grant of a Second Option Over the Nkana Smelter, Refinery and Acid Plant." Photocopy, ZCCM Archives, Ndola, February 24, 2000.

"Zambia Country Assistance Review." Operations Evaluation Department Report no. 15675. Washington, DC: The World Bank, 1996.

"Zambia Country Assistance Strategy." Country Department 3, Africa Region Report no. 27654-ZA. Washington, DC: The World Bank, 2004.

Zambia Episcopal Conference, Christian Council of Zambia, and Zambia Evangelical Fellowship. *Marxism, Humanism, and Christianity.* Lusaka: Teresianum Press, 1979.

"Zambia Needs to Turn to God—Chiluba." *The Post*, May 22, 2005, 2.

Zambia Privatisation Agency/Zambia Development Agency, http://www.zpa.org.zm/ (accessed January 7, 2010).

"Zambia Prospects for Sustainable Growth 1995-2005." Economic Management and Social Policy Group, Africa Region Report no.15477-ZA. Washington, DC: The World Bank, 1996.

"Zambia: The Challenge of Competitiveness and Diversification." Private Sector Unit, Africa Region Report no. 25388-ZA. Washington, DC: The World Bank, 2003.

"Zambia to Let Mopani Mine Operate for Now." *Creamer Media's Engineering News Online*, July 1, 2005. www.creamermedia.net/end/news/today/?show=69927 (accessed July 3, 2005).

"Zambianization: Steps That Led to a Big Success." *Mining Mirror*, October 22, 1984, 6-7.

Zambia's Mining Industry: The First 50 Years. Ndola: Roan Consolidated Mines, Ltd., 1978.

"ZCCM Copper Lower Than Estimates." *Mining Mirror*, February, 1997, 4.

ZCCM-Investment Holdings. http://www.zccm-ih.com.zm (accessed December 31, 2009).

"ZCCM Transformation Plan." Item 14.3.9E, ZCCM Archives, Ndola, September 16, 1998.

"ZESCO Delays Banana Project." *Sunday Times of Zambia*, December 5, 2004, 1.

♀

Index

About the Author

Elizabeth C. Parsons holds the M.Div. degree from Harvard Divinity School and the Ph.D. from the University of KwaZulu-Natal. She teaches courses in religion and development at Boston University School of Theology and is an affiliated faculty member of the University's African Studies Center.

www.ingramcontent.com/pod-product-compliance
Ingram Content Group UK Ltd.
Pitfield, Milton Keynes, MK11 3LW, UK
UKHW041631031026
15587UKWH00001B/215